BARRON'S
VISUAL LEARNING

Chemistry

Copyright © UniPress Books Limited 2021

Published by arrangement with UniPress Books Ltd

Publisher: Nigel Browning

Page design and illustrations: Lindsey Johns

Project manager: Natalia Price-Cabrera

Editorial consultant: Catherine Gefell

First edition published in North America by Kaplan, Inc.,
d/b/a Barron's Educational Series

All rights reserved. No part of this book may be reproduced
in any form or by any means without the written permission
of the copyright owner.

Published by Kaplan, Inc., d/b/a Barron's Educational Series

750 Third Avenue
New York, NY 10017

www.barronseduc.com

ISBN: 978-1-5062-8096-7

Kaplan, Inc., d/b/a Barron's Educational Series print books are
available at special quantity discounts to use for sales promotions,
employee premiums, or educational purposes. For more information
or to purchase books, please call the Simon & Schuster
special sales department at 866-506-1949.

Printed in China

10 9 8 7 6 5 4 3 2 1

Born and raised in Turkey, **Dr. Ali O. Sezer** is a Professor of Chemistry
at California University of Pennsylvania with a Ph.D. in Chemical and
Materials Engineering from the University of Nebraska-Lincoln. His expertise
is in the area of general, physical, and polymer chemistry. Dr. Sezer has been
actively involved in developing educational resources for chemistry including
two illustrated books entitled 30 Second Chemistry and Know-It-All Chemistry.
With many years of teaching experience, he believes in the power of science
education as an indispensable tool in creating a society that works for the
common good of humanity and a better human experience on
Earth where resources are limited.

BARRON'S

VISUAL LEARNING

Chemistry

AN ILLUSTRATED GUIDE FOR ALL AGES

DR. ALI O. SEZER

CONTENTS

iNTRODUCTiON

Chemistry studies matter, which is defined as anything in our universe with mass and volume. Through observations, experimentations, hypotheses, and theories, chemistry seeks to explain not only the properties of matter, such as its mass, color, smell, and density, but also how and why it undergoes transformation when its environmental conditions are changed.

Chemistry plays a central role in all branches of science, but it is also all around us—everything we see is made of chemicals, and everything we do on a daily basis involves some kind of chemistry process.

Without chemistry it would be impossible to understand how our physical world works, and yet we rarely (if ever) take the time to think about the impact chemistry has on our day-to-day lives.

On the following pages is a timeline of discovery that will go some way to explain the origins of what we call chemistry, together with flagging up major milestones that have occurred over centuries.

A Timeline of Discovery

The idea that the world is composed of tiny indivisible particles ("atomos") was first proposed in the 5th century B.C.E., by the ancient Greek philosopher, Leucippus, and his student, Democritus. However, the philosopher and polymath,

Aristotle (384–322 B.C.E.), believed that matter was continuous and infinitely divisible. Such was Aristotle's influence that Leucippus's and Democritus's theory would not receive widespread acceptance for another 2000 years.

From ancient times, the alchemical view of the world dominated scientific, philosophical, and theological discussions. Alchemy carried elements of science, philosophy, and mysticism, with alchemists striving to transmute ordinary metals into gold —the "perfect" metal—and discover the elixir of life. Alchemy remained popular until the late 17th century, when the emergence of forward-thinkers such as Robert Boyle (1627–1691) and later Antoine-Laurent de Lavoisier (1743–1794), and a more informed understanding of metallurgy, signaled the end of the alchemical era.

500 B.C.E.

Leucippus and Democritus proposed the particular nature of matter against popular belief in their time.

Yet alchemy still played a critical role in the emergence of modern science, with early scientists looking back on alchemical principles as they ventured into the atomic and particular view of matter. In 1661, Robert Boyle, an Anglo-Irish philosopher, chemist, and physicist, published *The Sceptical Chymist*, in which he shared his work on gases. He proposed that elements were made up of "corpuscles" (atoms) that could combine to give different chemical substances. Many other colorful personalities of the 17th century carried Boyle's work further, leading to the development of experimental chemistry and the discovery of many elements.

Antoine-Laurent de Lavoisier, a French chemist, carefully synthesized the knowledge accumulated before him to perfect the art of deriving theories from experimental observations. He studied the combustion reactions of various elements with oxygen and discovered that mass was conserved during a chemical reaction (the law of conservation of mass). He was also the first to write an extensive list of elements, as well as helping to construct the metric system and chemical nomenclature. Thought of as the father of modern chemistry, de Lavoisier opened the doors for those who succeeded him.

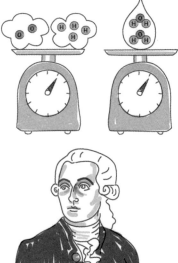

De Lavoisier demonstrated the law of conservation of mass by weighing compounds before and after a chemical reaction.

1789

1793

From numerous experimental observations, another French chemist, Joseph Louis Proust (1754–1826), brilliantly formulated the law of definite proportions, which stated that—regardless of the source or the method of preparation—each chemical compound always had the same elements in the same proportions.

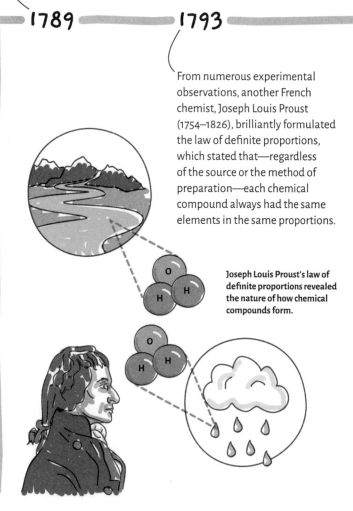

Joseph Louis Proust's law of definite proportions revealed the nature of how chemical compounds form.

Robert Boyle believed that elements consisted of corpuscles. Combined, these produced a variety of chemical substances.

1661

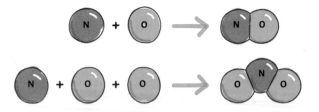

John Dalton (1766–1844), an English chemist, physicist, and meteorologist, attributed the law of definite proportions to the particulate nature of matter and postulated that the fixed elemental proportions in chemical compounds were due to the fact that matter was made of atoms. He further demonstrated that elements could also combine in different fixed ratios to produce different chemical compounds (the law of multiple proportions).

John Dalton explained that the formation of compounds in certain ratios was possible due to the atomic nature of matter.

1904

Thompson's "plum pudding" model of the atom.

1803 ================ 1895

Dalton's "solid sphere" model of the atom.

In 1803, Dalton published his atomic theory of matter, in which he defined atoms as indivisible solid spheres, which made all matter. His theory ignited a cascade of scientific studies, which led to a century of head-spinning scientific development, the discovery of more elements, and the construction of the first periodic table of the elements.

It took almost the entire 19th century for scientists to discover the existence of light outside the visible region the human eye can see. Wilhelm Conrad Röntgen (1879–1955), a German physicist, discovered an invisible form of light that he called "x-rays," which were capable of penetrating human flesh. In 1895, he published the first ever x-ray image (of his wife's left hand) and stunned the scientific community. In 1901 he was awarded the inaugural Nobel Prize in Physics for this discovery, which he gifted to humanity, refusing to receive even a penny as compensation. Unaware of the dangers of these invisible rays, he died of cancer, but not before opening the doors to a completely revolutionized medical imaging field.

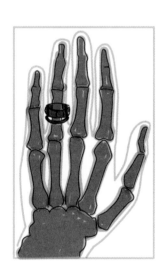

The first ever x-ray image showing Röntgen's wife's left hand.

1905

Albert Einstein elucidated the true nature of light and how it interacts with matter.

In 1905, Albert Einstein, a German theoretical physicist, proposed that light was composed of packets of energy—called photons—that travel in space in a wave form (electromagnetic radiation). An intense debate among scientists about the true nature of matter and how it interacts with light followed. This led to the formulation of the quantum theory, which provided the tools scientists needed to explore further, resulting in discoveries that would change the human experience on Earth forever.

The impact of chemistry on society can be seen starting in the early 1900s. In 1905, another German, the chemist Fritz Haber (1868–1934), invented a process that produced ammonia from the combustion of atmospheric hydrogen with nitrogen. This was a pivotal point in the history of agriculture, as ammonia is essential in the production of fertilizer. The agricultural world went on to experience a boom, providing much-needed food for people and animals.

The production of ammonia, essential for fertilizer, led to a boom in the agriculture industry.

1911

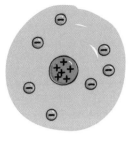

Rutherford's "planetary" model of the atom.

1913

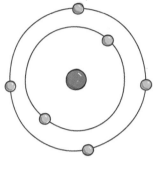

Bohr's "circular orbit" model of the atom.

1926

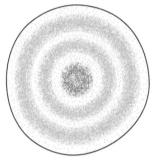

SchrÖdinger's "electron cloud" model of the atom.

1928

However, Sir Alexander Fleming's (1881–1955) serendipitous discovery of penicillin in 1928 is perhaps the quintessential scientific achievement: the Scottish researcher's discovery enabled the creation of the antibiotics that are still essential for eradicating many diseases even today.

The discovery of penicillin demonstrates the importance of science in society.

Not all scientific discoveries were immediately beneficial, though. When chemists first synthesized polyethylene in 1898, they saw no use for the sticky white substance. However, an industrial process was discovered (accidentally) in 1933 that could produce polyethylene on a large scale. This marked the beginning of the plastics era and the widespread use of what is now the most common plastic. Although environmental concerns have now changed many people's opinions of plastics, they were an invaluable commodity during the 20th century—especially during industrialization—and helped revolutionize daily life around the world.

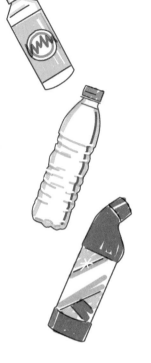

Plastics have impacted many areas of human life.

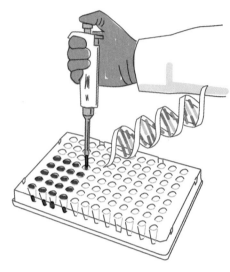

PCR testing is a standard in DNA chemistry.

1933 **1964** **1971**

In 1983, Kary Banks Mullis (1944–2019), an American biochemist, discovered the polymerase chain reaction (PCR) that established the foundations of the ultrasensitive PCR-test used today to identify viral and bacterial infections in biological specimens. This discovery revolutionized DNA chemistry and Mullis was granted the Nobel Prize in Chemistry for it in 1993. The PCR mechanism allows chemists to amplify or copy DNA samples hundreds of millions of times, making their detection fast and easy in various applications such as DNA fingerprinting, viral and bacterial detection, and the diagnosis of genetic disorders. The incredible impact PCR testing has shown through the HIV epidemic and the coronavirus pandemic is sufficient to appreciate the enormous significance of Mullis' discovery.

With a better understanding of the chemical properties of matter, as well as more advanced tools, new, specialized chemicals were created throughout the 20th century. Compounds with properties that are similar to both liquids and crystalline solids, for example, led to the invention of the liquid-crystal display (LCD) in 1964. This had an incredible impact on the electronics industry, so much so that it is hard to imagine life today without the monitors and screens we rely on so much.

In 1971, Dr. Akira Endo, a Japanese biochemist, discovered a class of chemical compounds called statins that were later shown to be effective in decreasing the level of low-density lipoproteins (LDL), commonly referred to as bad cholesterol, in the blood. This discovery is credited for saving tens of millions of people as high levels of cholesterol in blood is linked to serious health risks including coronary heart disease.

The invention of the liquid-crystal display in the mid-1960s revolutionized the electronics industry.

Statins prevent cholesterol build up in the arteries.

1990 ================================= **1998**

	114 **Fl** Flerovium

The 1990s were dominated by scientific research in the area of supramolecular chemistry, the study of new molecules that are 1100 nanometers in size, much larger and more complex than individual molecules. The excitement over supramolecules started with the 1987 Nobel Prize in Chemistry, awarded to Jean-Marie Lehn, Donald J. Cram, and Charles J. Pedersen. Macromolecules form as a result of self-assembly of molecular building blocks via noncovalent bonding to form nanometer-sized structures, and complex molecular systems are now deemed critically important in many applications such as drug development, chemical and biological sensors, nanoscience, molecular devices, and nanoreactors. Scientific research through the 1990s and early 2000s produced a second Nobel Prize in Chemistry in 2016 for contributions to the field of molecular machines built from supramolecules, invaluable in many different areas including the delivery of therapeutic cancer agents to targeted cells in the human body.

From 1998 onward, significant additions were made to the periodic table with the discovery of six new elements, atomic numbers 113 to 118, completing the seventh row of the table. Flerovium (1998), livermorium (2000), nihonium and moscovium (2003), oganesson (2006), and tennessine (2010) were discovered via artificial transmutation by bombarding heavy nuclei with beams of lighter nuclei in carefully controlled experiments carried out in particle accelerators. Since these elements only exist for a fraction of a second, their confirmation and acceptance by the scientific community took several years. It is certainly conceivable that future discoveries may add an eighth row to the periodic table.

2000 — 116 **Lv** Livermorium

2003 — 113 **Nh** Nihonium

115 **Mc** Moscovium

2006 — 118 **Og** Oganesson

2010 — 117 **Ts** Tennessine

Therapeutic drugs can be hosted in the cavities of three-dimensional assemblies created by supramolecules for targeted delivery.

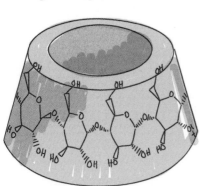

As a science, chemistry has played a central and crucial role in all of these developments, as well as countless others. However, it is impossible to fully appreciate what scientists have accomplished over the last 200 years without an understanding of the field of chemistry. That is where this book steps in. With the aid of powerful illustrations and fascinating real-life connections, *Visual Learning: Chemistry* builds from the atom out, exploring the fundamental principles of chemistry. In doing so, it will help you gain an understanding and appreciation of how chemistry is applied across a diverse range of scientific and technological fields, in turn shaping and defining our modern society and way of life.

CHAPTER 1

CHEMISTRY: A CENTRAL SCIENCE

Chemistry is a science concerned with matter, from its composition, structure, and changes, to its interactions with other matter and energy. From Democritus and Aristotle, through the experimental era of alchemy, to the father of modern chemistry, Antoine-Laurent de Lavoisier (1743–1794), chemistry has been around for a long time. During this time, scientists have used observation and experimentation to develop theories that try to better explain our physical world. This accumulated knowledge led to an incredible advancement in the field that started around the mid-19th century and has continued until today, positioning modern chemistry as a fundamental branch of science.

THE FUNDAMENTAL ROLE OF CHEMISTRY

The entire physical world is made up of the matter and energy that chemistry seeks to explore and explain, so it is easy to see just how important chemistry is in understanding the behavior of matter, from atoms to stars, from rocks to living things. Chemistry is, therefore, not only essential in itself, but it is also central to many other fields of scientific study.

Chemistry and the Human Experience

With a role in a wide range of applications, chemistry impacts human life in a variety of ways.

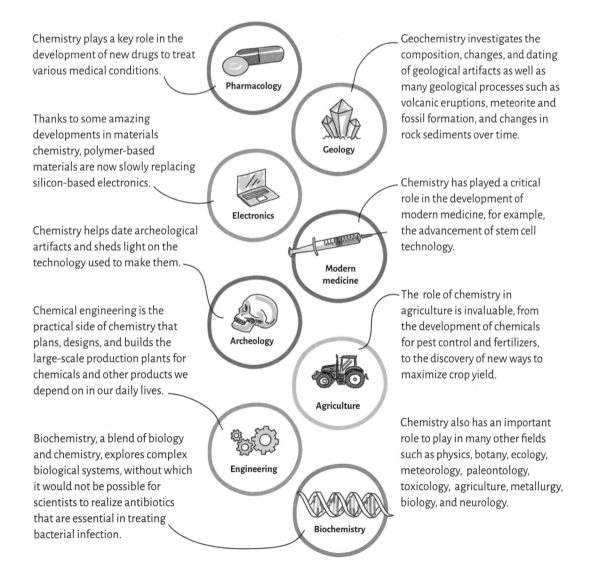

Chemistry plays a key role in the development of new drugs to treat various medical conditions.

Pharmacology

Geochemistry investigates the composition, changes, and dating of geological artifacts as well as many geological processes such as volcanic eruptions, meteorite and fossil formation, and changes in rock sediments over time.

Geology

Thanks to some amazing developments in materials chemistry, polymer-based materials are now slowly replacing silicon-based electronics.

Electronics

Chemistry has played a critical role in the development of modern medicine, for example, the advancement of stem cell technology.

Modern medicine

Chemistry helps date archeological artifacts and sheds light on the technology used to make them.

Archeology

The role of chemistry in agriculture is invaluable, from the development of chemicals for pest control and fertilizers, to the discovery of new ways to maximize crop yield.

Agriculture

Chemical engineering is the practical side of chemistry that plans, designs, and builds the large-scale production plants for chemicals and other products we depend on in our daily lives.

Engineering

Chemistry also has an important role to play in many other fields such as physics, botany, ecology, meteorology, paleontology, toxicology, agriculture, metallurgy, biology, and neurology.

Biochemistry, a blend of biology and chemistry, explores complex biological systems, without which it would not be possible for scientists to realize antibiotics that are essential in treating bacterial infection.

Biochemistry

Branches of Chemistry

There are five conventional branches of chemistry: physical, analytical, inorganic, organic, and biochemistry. These can be subdivided into more than 40 sub-branches, with an increasing number of specialized subdisciplines evolving as new and exciting application areas emerge. Despite its many sub-disciplines, the fundamental role of chemistry is the same: to study matter and the types of changes that occur during physical, chemical, and nuclear processes. However, each discipline explains chemistry processes from a different perspective, providing us with an indispensable tool for understanding the universe we live in.

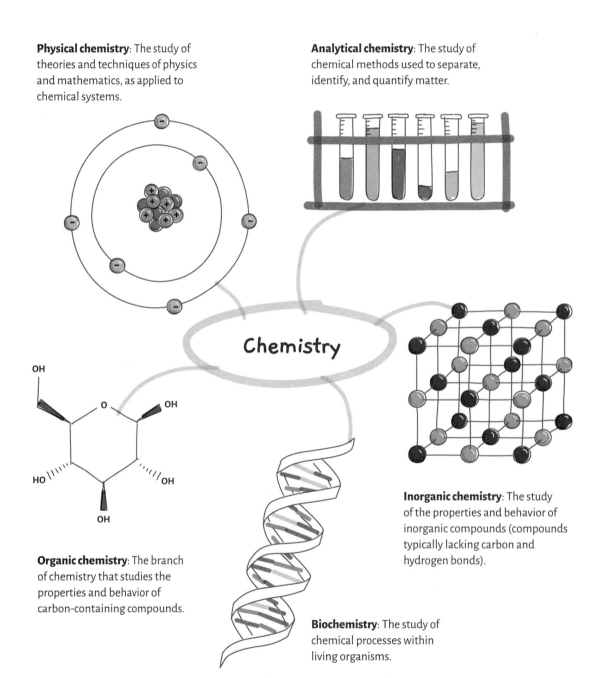

Physical chemistry: The study of theories and techniques of physics and mathematics, as applied to chemical systems.

Analytical chemistry: The study of chemical methods used to separate, identify, and quantify matter.

Organic chemistry: The branch of chemistry that studies the properties and behavior of carbon-containing compounds.

Inorganic chemistry: The study of the properties and behavior of inorganic compounds (compounds typically lacking carbon and hydrogen bonds).

Biochemistry: The study of chemical processes within living organisms.

Chemistry

MATTER

In general terms, matter—whether it is visible or invisible to the human eye—is anything that has mass and takes up volume. All physical objects are made of matter, which itself is composed of tiny building blocks known as **atoms**. There are three main physical forms of matter: **solid**, **liquid**, and **gas**.

Defining Matter

The **mass** of an object is a measure of the amount of matter it contains.

The **volume** of an object is a measure of the amount of space it occupies.

Density is a characteristic property of matter that is defined as the mass of a unit volume of material.

Density = mass/volume

MATTER

has → VOLUME

has → MASS

VOLUME is → Occupied space

VOLUME determines → Density: $d = \dfrac{Mass}{Volume}$

MASS is → Amount of substance

Density: characterizes → Material or substance

Low density — High density

States of Matter

The **solid** state is the highest density state of matter, with atoms or molecules packed closely together in fixed positions.

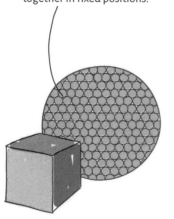

The **liquid** state has a density that is between the solid and gas states. In the liquid state, atoms or molecules can move around, but stay in close proximity to each other.

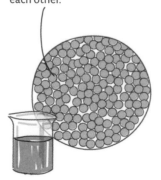

The lowest density belongs to the **gas** state of matter, with large separation between atoms or molecules. Gas particles move freely due to insignificant attraction between particles and low density.

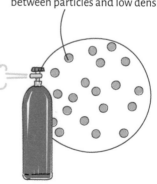

CLASSIFICATION OF MATTER

Matter is all around us. In our daily lives we come into contact with it in many different guises; we breathe matter, sit on it, wear it, drink it, and eat it. This matter can be found in either a **pure** form (made with pure substances) or as **mixtures** that consist of different substances.

Classification of Matter

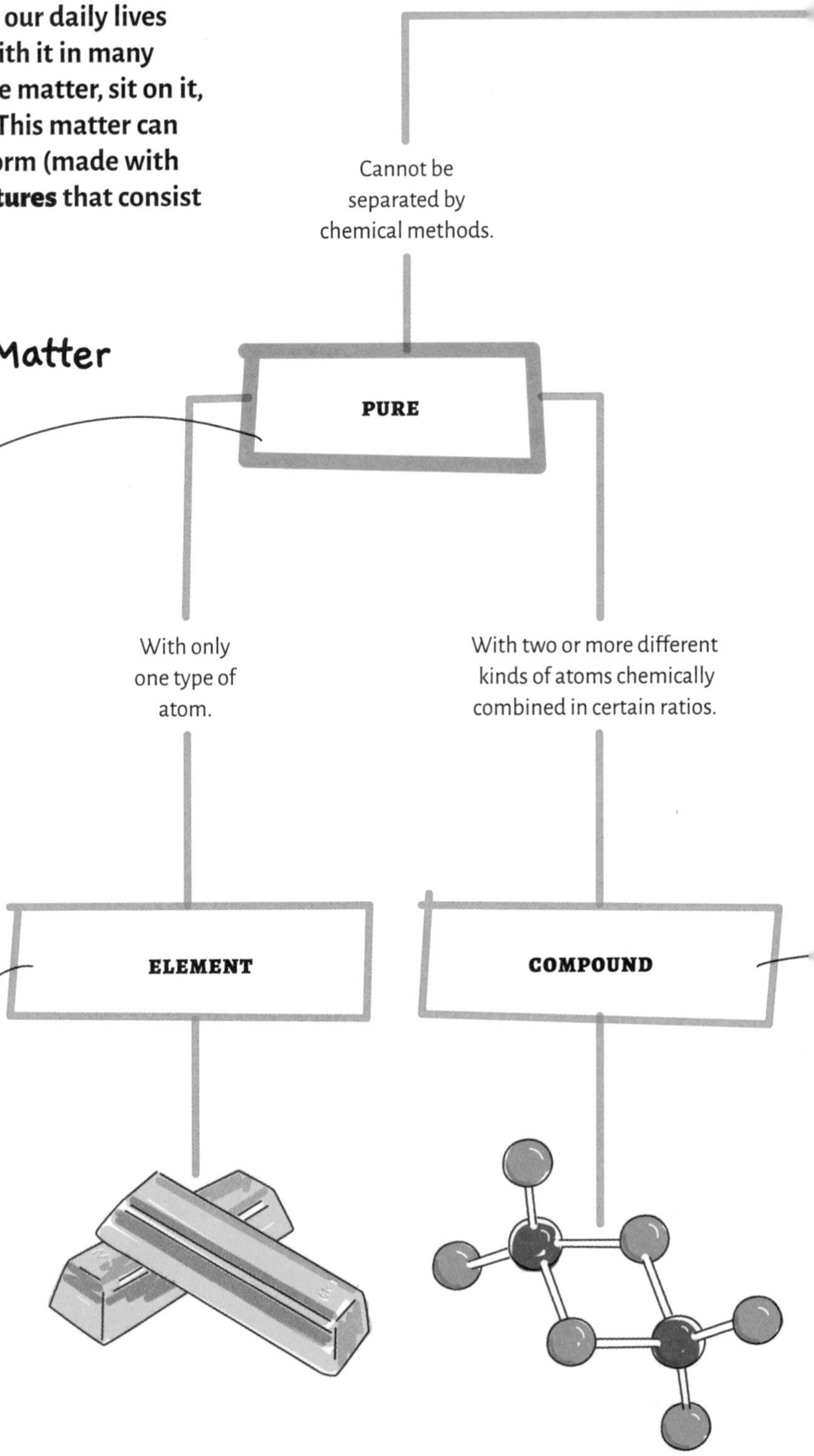

A **pure** form of matter is a substance that has a well-defined and fixed composition that does not change from sample to sample. A pure substance forms through the permanent combination of elements that require chemical processes to separate them. Table salt (NaCl), for example, is a pure substance consisting of chemically combined sodium and chlorine atoms, while water (H_2O) is a pure substance made up of two hydrogen atoms and one oxygen atom.

A pure substance can be either elemental or compound in form. An **element** is a fundamental substance consisting of only one type of atom, from which all material things are constructed. The element gold, for example, is made up solely of gold atoms.

Cannot be separated by chemical methods.

PURE

With only one type of atom.

With two or more different kinds of atoms chemically combined in certain ratios.

ELEMENT

COMPOUND

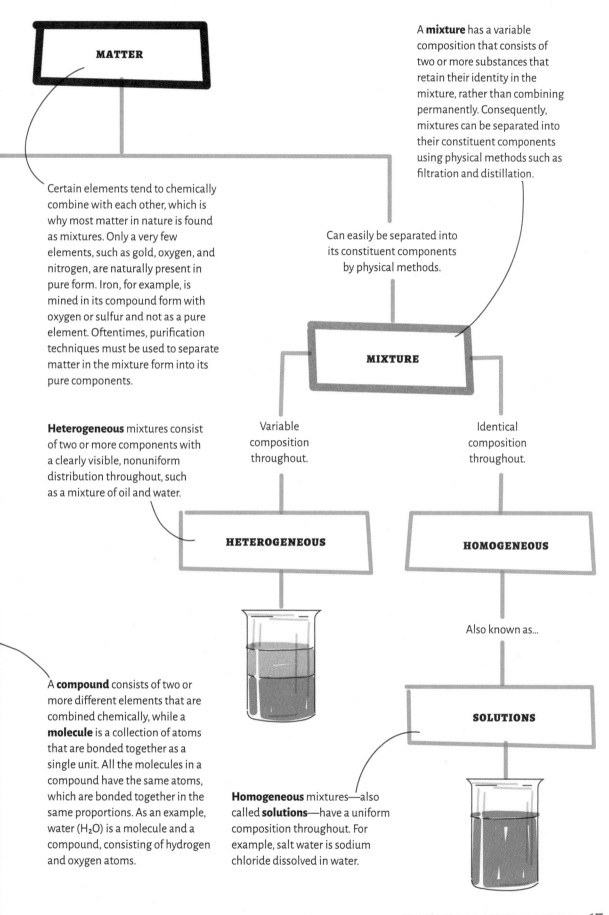

MATTER

A **mixture** has a variable composition that consists of two or more substances that retain their identity in the mixture, rather than combining permanently. Consequently, mixtures can be separated into their constituent components using physical methods such as filtration and distillation.

Certain elements tend to chemically combine with each other, which is why most matter in nature is found as mixtures. Only a very few elements, such as gold, oxygen, and nitrogen, are naturally present in pure form. Iron, for example, is mined in its compound form with oxygen or sulfur and not as a pure element. Oftentimes, purification techniques must be used to separate matter in the mixture form into its pure components.

Can easily be separated into its constituent components by physical methods.

MIXTURE

Variable composition throughout.

Identical composition throughout.

Heterogeneous mixtures consist of two or more components with a clearly visible, nonuniform distribution throughout, such as a mixture of oil and water.

HETEROGENEOUS

HOMOGENEOUS

Also known as...

A **compound** consists of two or more different elements that are combined chemically, while a **molecule** is a collection of atoms that are bonded together as a single unit. All the molecules in a compound have the same atoms, which are bonded together in the same proportions. As an example, water (H_2O) is a molecule and a compound, consisting of hydrogen and oxygen atoms.

Homogeneous mixtures—also called **solutions**—have a uniform composition throughout. For example, salt water is sodium chloride dissolved in water.

SOLUTIONS

PROPERTIES AND CHANGES OF MATTER

One of the most important roles of chemistry is to develop new and more useful materials that improve the human experience on Earth, such as changing chemical compounds to make pharmaceutical drugs that can treat various medical conditions. Developing new materials requires making changes to the **physical**, **chemical**, and/or **nuclear** properties of matter. These changes range from simple phase changes to permanent changes in the identity of the elements involved, but in all cases, the result is a new form of matter.

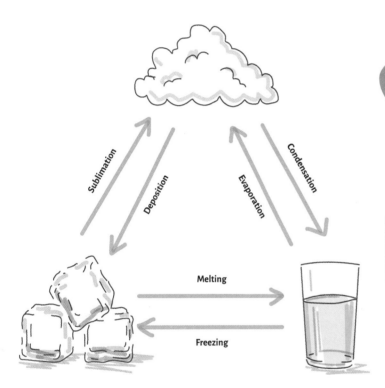

Sublimation

Deposition

Evaporation

Condensation

Changes in Matter

PHYSICAL CHANGE

Melting

Freezing

A **physical** property of a substance can be measured or observed without changing the identity of the substance, such as its odor, color, density, mass, and boiling temperature. Water melting from ice to liquid water, and liquid water vaporizing to steam, are both examples of **physical change**.

The energy involved in many changes in matter is often in the form of **heat** (thermal energy), which is energy on the move measured in joules (J) or calories (Cal). **Temperature** tells us the direction in which heat flows. Temperature is an indication of how hot or cold an object is, and heat naturally moves from a higher-temperature region to a lower-temperature region.

A **chemical** property is observed when a new substance is produced from the same elements in a different arrangement with different properties, such as iron rusting, a candle burning, and gasoline burning.

Burning wood is another example of a **chemical change**, as carbon and other elements in the wood are turned irreversibly into other compounds, which still consist of the same elements.

CHEMICAL CHANGE

$$^{2}_{1}\text{H} \quad + \quad ^{2}_{1}\text{H} \quad \longrightarrow \quad ^{4}_{2}\text{He} \quad + \text{energy}$$

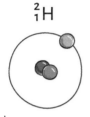
Deuterium

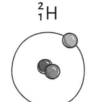
Deuterium

Helium

NUCLEAR CHANGE

Changes in **nuclear** properties involve changes to both the composition and identity of the atoms involved. For example, the **nuclear change** at the core of the sun involves two hydrogen atoms combining to form helium, releasing an enormous amount of energy.

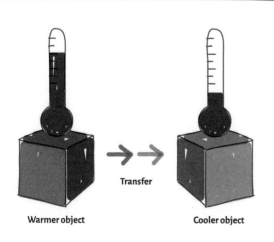

Transfer

Warmer object   Cooler object

Temperature is commonly expressed in Celsius (°C) or Kelvin (K).

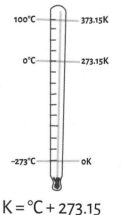

100°C — 373.15K

0°C — 273.15K

−273°C — 0K

$$K = {}^{\circ}C + 273.15$$

MATTER AND ENERGY

Energy can be defined as the ability to cause change. In other words, energy is required to make something happen that would not otherwise happen by itself. As it does not have volume or mass, energy is not classified as matter, but it can cause changes in matter. The law of conservation of energy states that the total energy of the universe is constant, which means energy cannot be created, nor can it be destroyed. It can, however, convert from one form to another.

Changes in matter are almost always accompanied by changes in energy. Small amounts of energy (0.5–45 kJ)/mole) are involved in physical changes, while chemical changes are typically associated with much larger energy transformations (200–900 kJ/mole). Energy in different forms can either be absorbed or released in chemical and physical processes, while nuclear changes release a tremendous amount of energy (1.0×10^8–2.0×10^{11} kJ/mole).

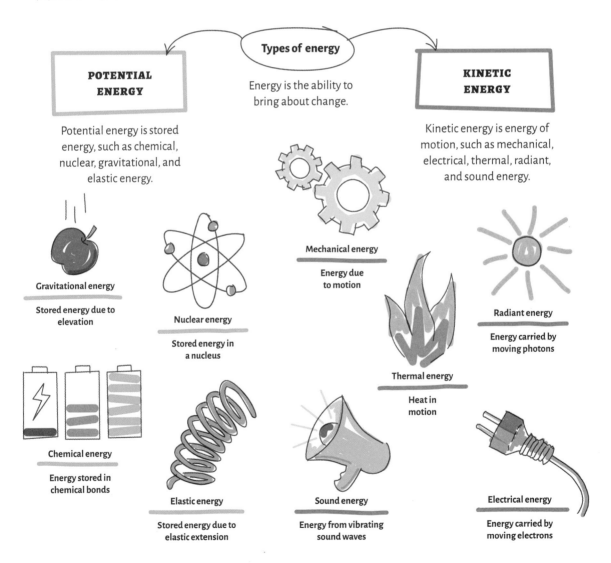

Types of energy

Energy is the ability to bring about change.

POTENTIAL ENERGY

Potential energy is stored energy, such as chemical, nuclear, gravitational, and elastic energy.

Gravitational energy

Stored energy due to elevation

Nuclear energy

Stored energy in a nucleus

Chemical energy

Energy stored in chemical bonds

Elastic energy

Stored energy due to elastic extension

Mechanical energy

Energy due to motion

Thermal energy

Heat in motion

Sound energy

Energy from vibrating sound waves

KINETIC ENERGY

Kinetic energy is energy of motion, such as mechanical, electrical, thermal, radiant, and sound energy.

Radiant energy

Energy carried by moving photons

Electrical energy

Energy carried by moving electrons

MEASUREMENT OF MATTER

Chemistry is a quantitative science, which means that observations and experiments involving the measurement of changes in matter are an essential part of what chemists do. Changes in matter are usually accompanied by changes in its properties, such as mass, volume, density, temperature, and composition. Scientists can measure the property of a substance and compare it with a standard having a known value of that property, and, in this way, their observations are both justifiable and repeatable.

Units of Measurement

Measurements in chemistry are expressed with a **numerical value** and a **unit** that indicates the standard against which the measured quantity is being compared.

The scientific system of measurement—the **International System of Units**, or **SI**—includes seven base units that are defined in terms of universal constants or properties that remain the same throughout the frame of reference. These fundamental quantities cannot be expressed in terms of other quantities, making them independent of any other units of measurement, including each other.

The fundamental quantities or units can be combined to define **derived** quantities or units. For example, speed is expressed in terms of length and time, as in miles per hour and meters per second.

273.15 K

Number   Unit

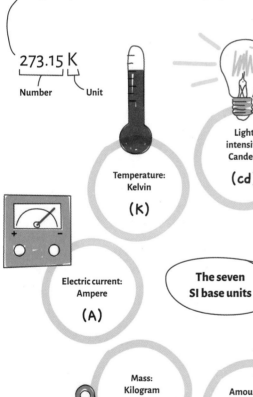

Temperature: Kelvin

(K)

Light intensity: Candela

(cd)

Length: Meter

(m)

Electric current: Ampere

(A)

The seven SI base units

Time: Second

(s)

Derived unit

Mass: Kilogram

(kg)

Amount of substance: Mole

(mol)

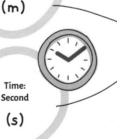

$$\text{Speed} = \frac{\text{length}}{\text{time}} = \frac{m}{s}$$

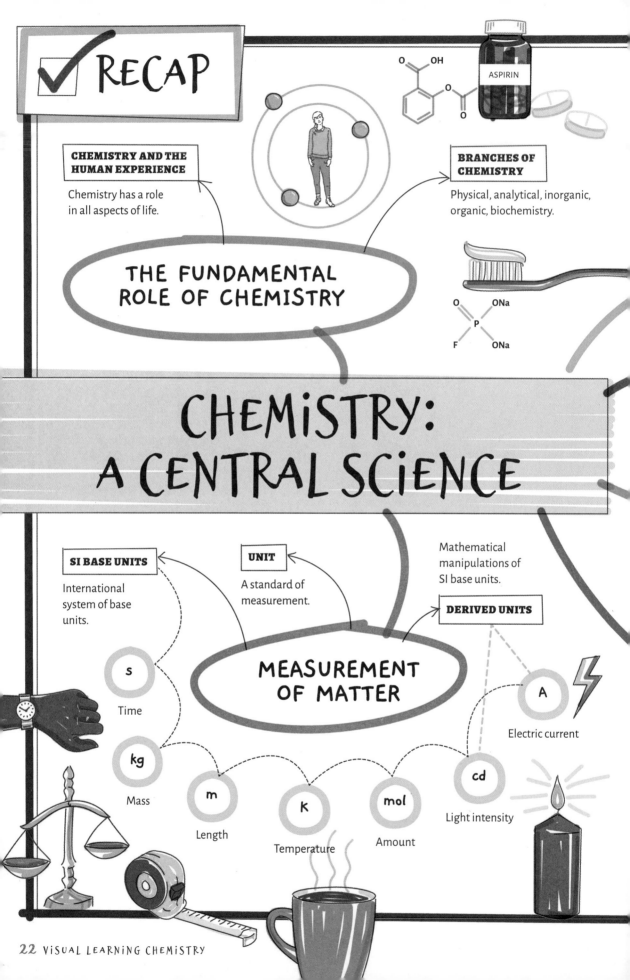

✓ RECAP

ASPIRIN

CHEMISTRY AND THE HUMAN EXPERIENCE

Chemistry has a role in all aspects of life.

BRANCHES OF CHEMISTRY

Physical, analytical, inorganic, organic, biochemistry.

THE FUNDAMENTAL ROLE OF CHEMISTRY

CHEMISTRY: A CENTRAL SCIENCE

SI BASE UNITS

International system of base units.

UNIT

A standard of measurement.

Mathematical manipulations of SI base units.

DERIVED UNITS

MEASUREMENT OF MATTER

s Time

kg Mass

m Length

k Temperature

mol Amount

A Electric current

cd Light intensity

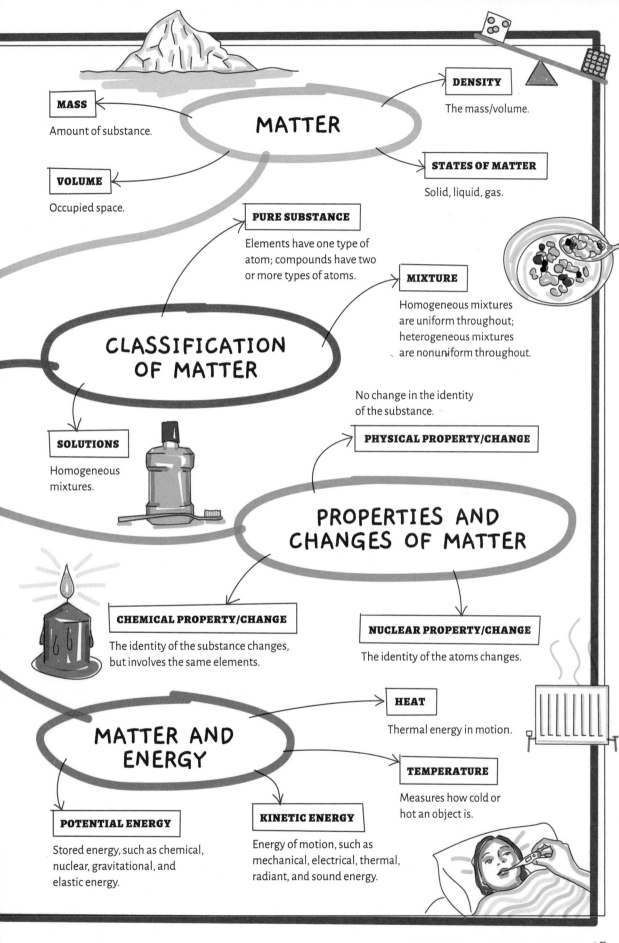

MASS

Amount of substance.

MATTER

DENSITY

The mass/volume.

VOLUME

Occupied space.

STATES OF MATTER

Solid, liquid, gas.

PURE SUBSTANCE

Elements have one type of atom; compounds have two or more types of atoms.

MIXTURE

Homogeneous mixtures are uniform throughout; heterogeneous mixtures are nonuniform throughout.

CLASSIFICATION OF MATTER

SOLUTIONS

Homogeneous mixtures.

No change in the identity of the substance.

PHYSICAL PROPERTY/CHANGE

PROPERTIES AND CHANGES OF MATTER

CHEMICAL PROPERTY/CHANGE

The identity of the substance changes, but involves the same elements.

NUCLEAR PROPERTY/CHANGE

The identity of the atoms changes.

HEAT

Thermal energy in motion.

MATTER AND ENERGY

TEMPERATURE

Measures how cold or hot an object is.

POTENTIAL ENERGY

Stored energy, such as chemical, nuclear, gravitational, and elastic energy.

KINETIC ENERGY

Energy of motion, such as mechanical, electrical, thermal, radiant, and sound energy.

CHAPTER 2

THE ATOM

Atoms, retaining all the chemical properties of elements, are the basic building blocks of matter in our natural world. All matter is built from one or more of 92 known elements. They are like the letters used to construct words, and from there a language, so understanding the atom and its components is crucial if you want to understand matter and how substances behave in our universe. Although the general structure of all atoms is similar, they are composed of different fundamental subatomic particles that makes each element unique with its own type of atoms and properties.

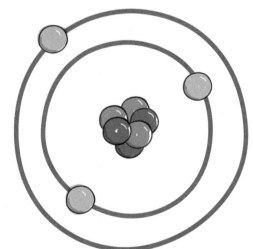

EVOLUTION OF ATOMIC MODELS

nfluenced by Aristotle, the ancient Greeks believed that all matter consisted of differing amounts of four basic substances: earth, fire, water, and air. In the 5th century B.C.E., Leucippus and his student, Democritus, proposed that matter possessed a "particular" nature, like a pool of water being separated into small drops. Each drop could be split into smaller and smaller drops until they become too small to see. Leucippus suggested that there must ultimately be tiny particles of water that could not be split into smaller entities. Democritus called those tiny particles "**atomos**" (which means "cannot be cut" in Greek), from which we derive the word "atom."

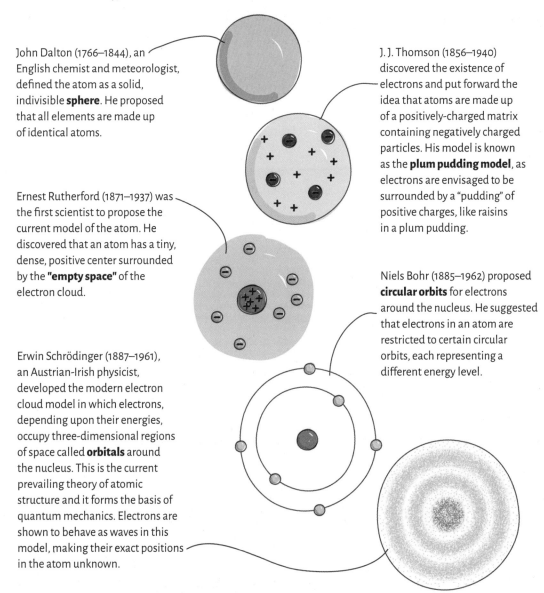

John Dalton (1766–1844), an English chemist and meteorologist, defined the atom as a solid, indivisible **sphere**. He proposed that all elements are made up of identical atoms.

Ernest Rutherford (1871–1937) was the first scientist to propose the current model of the atom. He discovered that an atom has a tiny, dense, positive center surrounded by the **"empty space"** of the electron cloud.

Erwin Schrödinger (1887–1961), an Austrian-Irish physicist, developed the modern electron cloud model in which electrons, depending upon their energies, occupy three-dimensional regions of space called **orbitals** around the nucleus. This is the current prevailing theory of atomic structure and it forms the basis of quantum mechanics. Electrons are shown to behave as waves in this model, making their exact positions in the atom unknown.

J. J. Thomson (1856–1940) discovered the existence of electrons and put forward the idea that atoms are made up of a positively-charged matrix containing negatively charged particles. His model is known as the **plum pudding model**, as electrons are envisaged to be surrounded by a "pudding" of positive charges, like raisins in a plum pudding.

Niels Bohr (1885–1962) proposed **circular orbits** for electrons around the nucleus. He suggested that electrons in an atom are restricted to certain circular orbits, each representing a different energy level.

SCiENTiFiC LAWS OF CHEMiCAL COMBiNATION

The atomic view of matter remained elusive until around the start of the 18th century, at which point a series of scientific observations (largely self-funded as hobbies) by some colorful personalities led to a better understanding of the atomic and particular nature of matter. The chemical properties of newly discovered elements were investigated by carrying out chemical reactions, with before and after reaction observations shared with the public. The accumulation of knowledge about elements and their chemical properties culminated in the development of important scientific laws on the chemical combination of elements.

The law of conservation of mass states that matter cannot be created or destroyed in a chemical reaction. For example, when a certain mass of hydrogen and oxygen reacts to form water, the mass of the water formed by the reaction is equal to the total mass of the reacting hydrogen and oxygen.

The law of definite proportions states that a chemical compound, such as water (H_2O), contains the same elements in exactly the same proportions by mass, regardless of the size of the sample or the source of the compound. So, whether it is from a river, rain, or a faucet, water always has two atoms of hydrogen and one atom of oxygen, combined by mass in the same ratio of 1:8.

The law of multiple proportions states that if two elements combine in different ratios, the resulting compounds are different. One nitrogen and one oxygen atom yield NO, but one nitrogen and two oxygen atoms yield NO_2—two different compounds from the same elements.

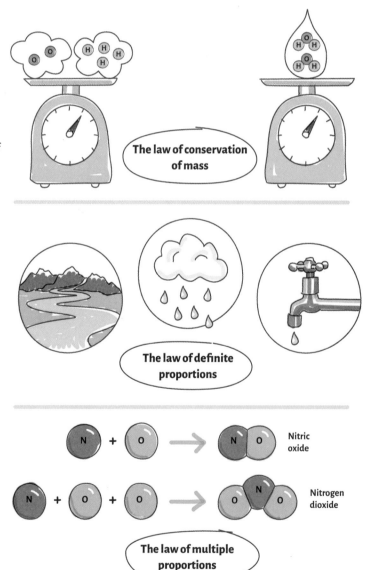

The law of conservation of mass

The law of definite proportions

Nitric oxide

Nitrogen dioxide

The law of multiple proportions

ATOMIC STRUCTURE

John Dalton developed the first modern atomic theory after a careful examination and interpretation of earlier work, especially the scientific laws of chemical combination. He proposed that all elements were made up of a single, unique type of atom, which could combine by chemical means to form compounds. Dalton clearly explained the difference between elements and compounds, and two of his four proposed ideas are still valid today, without modification. However, his ideas that an atom is indivisible and that all atoms in a given element are identical in every aspect have been revised, as they were proven incorrect.

Although Dalton's atomic theory explained earlier experiments, it did not provide details about the structure of the atom itself. Following his theory, however, several crucial discoveries were made that provided such details.

The modern model of the atom consists of a **nucleus** in which **nucleons** (neutrons and protons) are located, essentially making up the mass of the atom. **Electrons** are in the largely empty space (known as the **electron cloud**) outside the nucleus, which means most of the volume of an atom is empty space.

Neutral atoms have an equal number of electrons and protons, so they are electrically neutral. The lithium atom depicted here, for example, has three electrons and three protons.

Electron shells have different energies based on their distance from the nucleus. The energy of the electron increases as its distance from the nucleus increases.

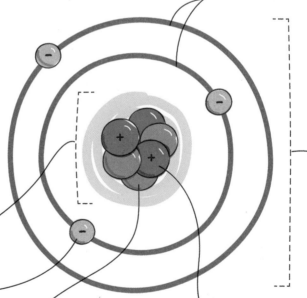

The size of the nucleus is ~ 10^{-14} m

The size of a typical atom is ~ 10^{-10} m

Electrons are negatively charged subatomic particles located outside the nucleus:
Relative charge = –1
Mass = 0.000549 amu
Size: ~ 10^{-18} m

Neutrons are subatomic particles with a neutral charge located in the nucleus:
Relative charge = 0
Mass = 1.00866 amu
Size: ~ 10^{-15} m

Protons are positively charged subatomic particles located in the nucleus:
Relative charge = +1
Mass = 1.00728 amu
Size: ~ 10^{-15} m

MENDELEEV AND THE PERIODIC TABLE

In 1869, a Russian chemist and inventor, Dmitri Ivanovich Mendeleev (1834–1907), proposed the periodic law that stated that the physical and chemical properties of all elements were a periodic function of their atomic masses. In other words, Mendeleev identified the mass of an element as the primary parameter upon which all other properties depended. At the time, there were 63 known elements, and Mendeleev's periodic table listed them based on their atomic weight. The modern periodic table uses the number of protons in an element's atom—the atomic number—as the primary factor in predicting elemental properties.

Mendeleev's Periodic Table

		Ti 50	Zr.........90	? 180
		V 51	Nb 94	Ta....... 182
		Cr 52	Mo 96	W 186
		Mn 55	Rh ...104.4	Pt 197.4
		Fe 56	Ru ...104.4	Ir 198
		Ni, Co... 59	Pd ...106.6	Os 199
H.............1		Cu 63.4	Ag......108	Hg..... 200
Be 9.4	Mg 24	Zn 65.2	Cd 112	
B 11	Al 27.4	? 68	Ur 116	Au 197
C 12	Si 28	? 70	Sn 118	
N.............4	P 31	As 75	Sb 122	Bi 210
O...........16	S 32	Se......79.4	Te....... 128 ?	
F........... 19	Cl 35.5	Br.........80	I 127	
Na........ 23	K 39	Rb 85.4	Cs........133	Tl 204
	Ca 40	Sr 87.6	Ba 137	Pb 207
	? 45	Ce 92		
	? Er....... 56	La.........94		
	? Y 60	Di......... 95		
	? In 75.6	Th 118		

Mendeleev's first periodic table listed elements based on their atomic weight.

The periodic law successfully predicted the weight and properties of eight undiscovered elements.

The behavior of some of the elements did not follow Mendeleev's prediction, so the chemist proposed that their atomic weights had been measured incorrectly. Science later confirmed that this was indeed the case.

Dmitri Mendeleev devised the first periodic table of elements.

The Modern Periodic Table

The modern periodic table lists 118 elements based on their number of protons, or **atomic number**. Scientific developments after Mendeleev concluded that the atomic number should be the fundamental parameter that best describes the properties of an element.

Although there are many variations of the periodic table of elements, each box gives an element's **symbol**, **name**, **atomic number**, and **atomic mass**.

Atomic number: The unique number of protons that defines the element.

Symbol of element: The official abbreviation for the name of the element.

Name of element: The official name of the element.

Atomic mass: The weighted average of all known isotopes of the element, representing the mass of one atom in units of atomic mass units (1 amu = 1.6625 × 10⁻²⁷ kg).

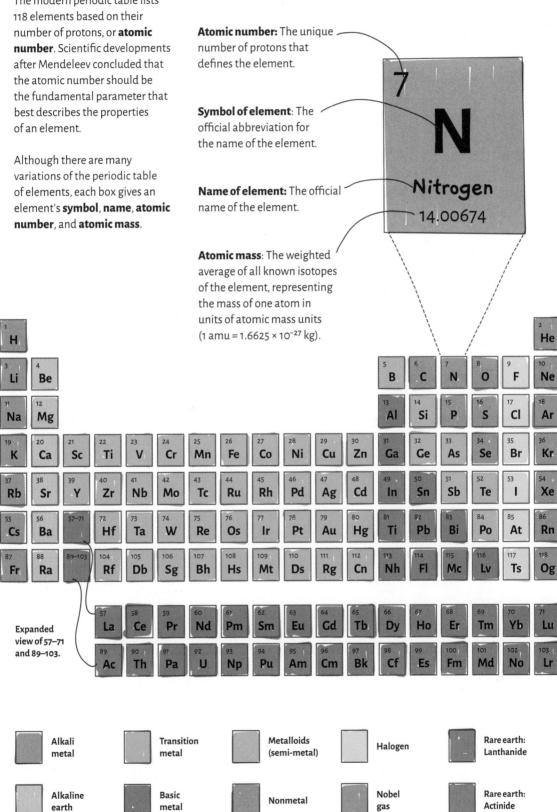

Expanded view of 57–71 and 89–103.

| Alkali metal | Transition metal | Metalloids (semi-metal) | Halogen | Rare earth: Lanthanide |
| Alkaline earth | Basic metal | Nonmetal | Nobel gas | Rare earth: Actinide |

iONS AND iSOTOPES

When an atom has an equal number of protons and electrons, it has no overall charge (positive or negative). However, when an atom has an unequal number of protons and electrons, it is charged and becomes an **ion**. The atoms in an element can also have the same number of protons and electrons, but a different number of neutrons, in which case they are called **isotopes**.

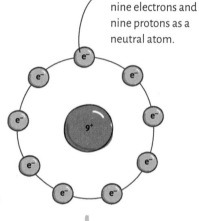

Fluorine (F) has nine electrons and nine protons as a neutral atom.

Ions

When the number of protons (+1 charge) is greater than the number of electrons (–1 charge), the atom has an overall positive charge. This positively charged ion is called a **cation**.

When the number of protons (+1 charge) is smaller than the number of electrons (–1 charge), the atom has an overall negative charge. This negatively charged ion is called an **anion**.

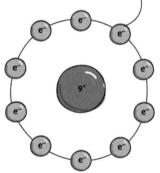

The fluoride ion (F⁻) has nine protons and ten electrons.

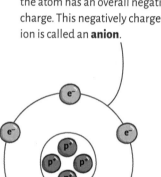

Isotopes

As they have the same number of protons and electrons, but different numbers of neutrons, isotopes have different masses. Hydrogen, for example, has three isotopes, which differ in the number of neutrons they possess. There are many examples of naturally occurring isotopes, both stable and unstable.

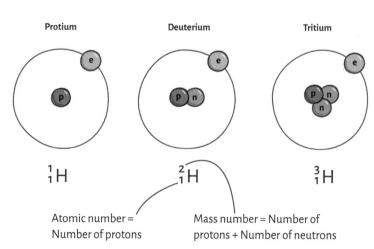

Protium

Deuterium

Tritium

$^{1}_{1}H$

$^{2}_{1}H$

$^{3}_{1}H$

Atomic number = Number of protons

Mass number = Number of protons + Number of neutrons

THE MOLE AND MOLAR MASS

Chemists use the fundamental SI unit **mole (mol)** to count atoms, ions, and molecules that are too small to handle in a laboratory setting. A mole is defined as the amount of substance that contains the same number of particles as there are atoms of a carbon-12 isotope in exactly 12 grams of carbon-12. This number is 6.022×10^{23} and is called **Avogadro's number**. The mass, in grams, of one mole of a substance is called the **molar mass**, which is expressed as **g/mol**.

The mass of a single atom is reported in atomic mass units. This is a very small mass because these particles are tiny. The mass of a molecule is determined by adding together the atomic masses from the periodic table.

In a laboratory setting, a mole of a substance containing 6.022×10^{23} of those particles is used instead of individual particles. Each pure substance has a unique molar mass, which gives the exact number of particles in a given sample. For example, the molar mass of water is 18.016 g/mol.

1 mol of water

Mass: 18.016g

The mass of a single oxygen atom is 16 amu.

The mass of a single hydrogen atom is 1.008 amu, so two hydrogen atoms weigh 2.016 amu.

$\times 6.022 \times 10^{23}$

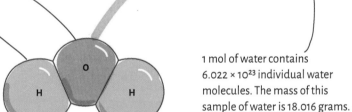

The mass of a single water molecule (H_2O) is the sum of the atomic masses it contains: 18.016 amu.

1 mol of water contains 6.022×10^{23} individual water molecules. The mass of this sample of water is 18.016 grams.

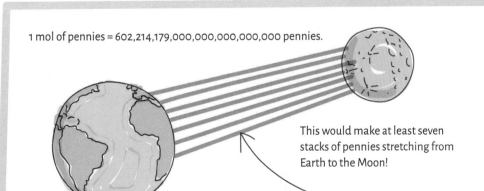

1 mol of pennies = 602,214,179,000,000,000,000,000 pennies.

This would make at least seven stacks of pennies stretching from Earth to the Moon!

PLANETARY

Rutherford (1911).

CIRCULAR ORBITS

Bohr (1913).

PLUM PUDDING

Thomson (1904).

SOLID SPHERE

Dalton (1803).

EVOLUTION OF ATOMIC MODELS

ELECTRON CLOUD

Schrödinger (1926).

THE ATOM

THE MOLE AND MOLAR MASS

AVOGADRO'S NUMBER

This is 6.022×10^{23}.

MOLAR MASS

The mass, in grams, of one mole of pure substance.

Provided the first periodic table (1869) and predicted the existence of elements.

DMITRI MENDELEEV

A table of 118 known elements arranged by their atomic number.

MOLE

A counting unit equal to Avogadro's number.

THE PERIODIC TABLE

Defined by Antoine-Laurent de Lavoisier (1774): mass cannot be destroyed or created.

THE LAW OF CONSERVATION OF MASS

THE LAW OF DEFINITE PROPORTIONS

Atoms combine in certain ratios to form compounds.

SCIENTIFIC LAWS OF CHEMICAL COMBINATION

THE LAW OF MULTIPLE PROPORTIONS

Atoms combine in different ratios to form different compounds.

ELECTRON SHELLS

Energy levels where electrons are located.

 ELECTRONS

Negatively charged subatomic particles located outside the nucleus.

THE ATOMIC STRUCTURE

 PROTONS

Positively charged subatomic particles in the nucleus.

 NEUTRONS

Neutral subatomic particles in the nucleus.

 NUCLEUS

Houses protons and neutrons.

 CATIONS

Positively charged atoms.

IONS AND ISOTOPES

The number of protons in an atom.

ATOMIC NUMBER

 ANIONS

Negatively charged atoms.

ISOTOPES

Atoms of the same element with different numbers of neutrons.

MENDELEEV AND THE PERIODIC TABLE

 **PERIOD**

Horizontal rows in the periodic table.

GROUP

Vertical columns in the periodic table.

CHAPTER 3

NUCLEAR CHEMISTRY

Nuclear chemistry is concerned with changes in the nucleus of an atom, which can be brought about by the interaction of two nuclei or by the impact of a subatomic particle on the nucleus. This branch of chemistry often involves studies of nuclear particles, nuclear forces, and nuclear reactions, exploring changes in the nucleus that can occur both naturally and artificially: radioactivity involves nuclear changes in naturally occurring unstable isotopes, while artificial transmutation refers to changes in the nucleus when it collides with a high-speed particle.

THE NUCLEUS

The nucleus of an atom occupies a tiny fraction of the atomic volume, but houses the characteristic number of protons and neutrons for a specific isotope. The electrostatic repulsion between the positively charged protons would normally be expected to break up the nucleus, but this does not happen because a strong **nuclear force** keeps the protons and neutrons together. The proton-to-neutron ratio in a nucleus has a direct effect on the strength of the nuclear force, determining whether a given nucleus is **stable** or **unstable**.

Nuclear Force

Nuclear force is a very strong attractive force that acts between subatomic particles that are extremely close together like protons and neutrons. The nuclear force in an atom overcomes the electrostatic repulsion forces between protons, keeping the nucleus and therefore the atom intact.

Nuclear force, strongest in nature, is identical for all nucleons. It does not matter if it is a neutron or a proton.

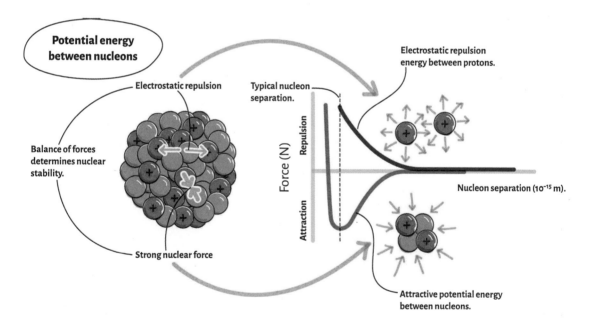

Potential energy between nucleons

Electrostatic repulsion

Typical nucleon separation.

Electrostatic repulsion energy between protons.

Balance of forces determines nuclear stability.

Strong nuclear force

Force (N)

Repulsion

Attraction

Nucleon separation (10^{-15} m).

Attractive potential energy between nucleons.

The nucleons come closer to each other until a certain distance allowed by nuclear forces. The attractive potential energy brings nucleons closer. At an equilibrium nucleon separation, the repulsive forces start to dominate. The repulsive force, therefore, determines the size of the nucleus. Nuclear force is only present between particles at a distance up to 1.0×10^{-15} m.

Nuclear force is what keeps electrons floating around the empty space around the nucleus. Its presence and enormous strength are the reasons we can generate tremendous amount of energy in nuclear power plants.

Nuclear Stability

Whether a nucleus is stable or unstable depends on the balance of forces within the nucleus, which is related to the ratio of protons and neutrons

Unstable isotopes tend to **decay** spontaneously, emitting **radiation** as they form a new and more stable atom.

The new atom emerging from the radioactive decay reaction typically has a new proton-to-neutron **ratio**, making it more stable.

An unstable isotope is radioactive, and is called a **radioisotope**.

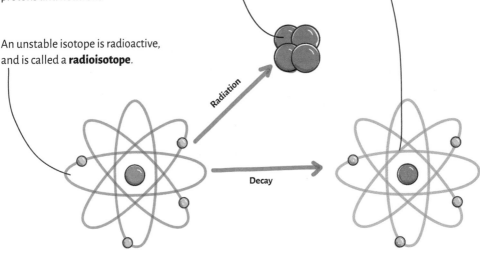

Radiation

Decay

Out of thousands of naturally occurring isotopes, only approximately 250 are stable with atomic numbers between 1 and 83. These stable isotopes form what is called the **belt of stability**.

Above the belt of stability, there are too many neutrons in the nucleus, making it unstable. Such isotopes undergo nuclear decay to form more stable nuclei by emitting beta particles.

Below the belt of stability, the nucleus is unstable because it has too many protons. Such isotopes undergo nuclear decay to form more stable atoms by emitting positrons or capturing electrons.

All isotopes with atomic numbers higher than 83 are unstable and they undergo radioactive decay to become more stable by emitting alpha particles.

All isotopes with an equal number of protons and neutrons are stable.

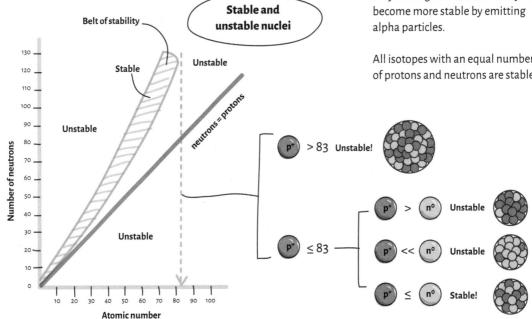

Stable and unstable nuclei

Belt of stability

Stable

Unstable

neutrons = protons

Number of neutrons

Unstable

Unstable

Atomic number

p^+ > 83 Unstable!

p^+ ≤ 83

p^+ > n^o Unstable

p^+ << n^o Unstable

p^+ ≤ n^o Stable!

Nuclear Binding Energy

Nuclear binding energy (E_b) is the energy that is needed to separate a nucleus into its protons and neutrons; it is high for stable nuclei and low for unstable nuclei.

Nucleus with smaller mass.

Binding energy

$$E_b = \Delta m \times c^2$$

Separated nucleons with greater mass.

The mass of a nucleus is always less than the total mass of the protons and neutrons it contains. This is called the **mass defect** (Δm). This difference in mass is converted into binding energy, as determined by Einstein's mass-energy relationship involving the speed of light ($c = 2.998 \times 10^8$ m/s).

Lithium−7

7.016005 amu

7.05658 amu

Mass defect is a negative number, as it represents the lost mass that is converted to energy.

$$\Delta m = 7.016005 - 7.05658 = -0.040575 \text{ amu}$$

Nuclear binding energy per nucleon

Nuclear binding energy naturally depends on the number of nucleons, and it generally increases for nuclei with low mass numbers because, at low mass numbers, the attractive nuclear forces dominate the electrostatic repulsive forces between protons.

^{56}Fe   Most stable nucleus in nature

Area of very stable nuclei

Binding energy per nucleon

Number of nucleons (mass number)

Nuclear binding energy peaks near iron (Fe) with a mass number of 56, and then it tapers off for larger mass numbers because, at large mass numbers, the electrostatic repulsion forces between protons are dominant.

The most stable nucleus in nature is iron because of the largest nuclear binding energy among all known isotopes, which is why the nuclear explosions in the core of stars end with the production of iron.

Isotopes with lower mass numbers than iron tend to combine in nuclear fusion reactions to release enormous amounts of energy because attractive nuclear forces dominate in their nuclei.

Isotopes with higher mass numbers than iron tend to break up in nuclear fission reactions releasing tremendous amounts of energy because electrostatic repulsions among protons strongly dominate in their nuclei.

NUCLEAR CHANGES

The number of protons and neutrons in the nucleus changes during nuclear reactions. When this happens a new isotope of the same atom, or a completely new element, is created. These changes are called **nuclear transmutations** and can occur naturally through **radioactivity** or artificially in a particle collider setting under carefully controlled conditions.

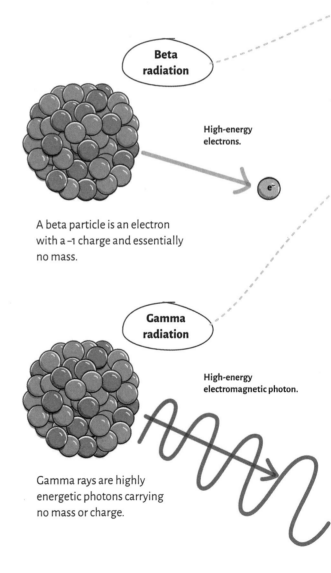

Alpha radiation

Two protons and two neutrons.

An alpha particle is essentially a helium-4 isotope with two protons and two neutrons.

Natural Transmutation

Natural transmutation, also called radioactivity, is the spontaneous decay of an unstable, naturally occurring isotope. Radioisotopes emit **alpha (α)**, **beta (β)**, and **gamma (γ)** radiation in decay reactions to form new and more stable elements or isotopes.

Radiation exposure is a part of everyday life. It comes from the food we eat, the air we breathe, and the environment we live in. Some of it comes from natural sources (alpha, beta, and gamma radiation) and while other forms of radiation exposure occur due to human activity such as medical imaging, diagnoses, and treatment.

Natural radiation exposure can take place inside the human body as well through the food we eat. This type of exposure can be particularly dangerous as it causes tissue damage of internal organs.

Beta radiation

High-energy electrons.

A beta particle is an electron with a –1 charge and essentially no mass.

Gamma radiation

High-energy electromagnetic photon.

Gamma rays are highly energetic photons carrying no mass or charge.

Radiation is a form of energy. Gamma radiation is the highest energy form of electromagnetic energy and has the most penetrating power. Beta radiation has less penetrating power and alpha particles have the least penetrating power of the three forms of radioactive emissions. Gamma rays and beta particles can easily penetrate the human body.

The second leading cause of lung cancer in the U.S.A. involves exposure to the naturally occurring radon-222 isotope (Rn-222), which comes from rocks and soil. Radon is frequently found in basements of buildings because it is heavier than air. It breaks down into polonium-218 (Po-218) and alpha particles within hours. Breathing air contaminated with Rd-222 causes internal exposure to alpha particles capable of causing tissue damage, especially in the lungs.

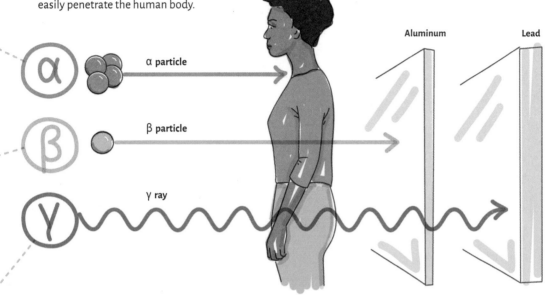

Artificial Transmutation

Artificial transmutation is not spontaneous. Instead, it needs to be induced by bombarding a nucleus with a high-energy particle using special equipment such as a particle accelerator. In 1919, Ernest Rutherford was the first scientist to successfully carry out an artificial transmutation. He bombarded a nitrogen-14 isotope with alpha particles, transmuting it into oxygen-17 and hydrogen-1 isotopes.

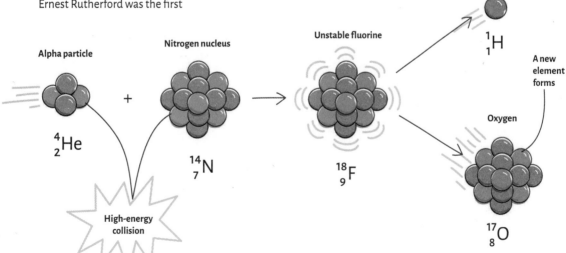

NUCLEAR EQUATIONS

Nuclear equations are the written form of nuclear reactions. In nuclear reactions, a **parent** nucleus either emits (**decay reaction**) or absorbs (**capture reaction**) a particle to turn into a daughter nucleus. In **fusion reactions,** two lighter parent atomic nuclei collide and fuse to give birth to a heavier "daughter" nucleus. In **fission reactions**, a heavier parent turns into two or more lighter daughter nuclei. Fission and fusion reactions are accompanied by emissions of nuclear particles and/or gamma radiation. In a nuclear equation, atomic mass and atomic number must be conserved.

Radiation Types

There are several different types of particles and radiation commonly involved in nuclear reactions, which are either emitted or captured by parent nuclei.

	Alpha particles: helium nuclei ejected from the nucleus of an atom	$^4_2\alpha = {}^4_2He$
	Beta particles: electrons ejected from the nucleus of an atom	$^0_{-1}\beta = {}^0_{-1}e$
	Positron particles: ejected from the nucleus of an atom	$^0_{+1}\beta = {}^0_{+1}e$
	Proton particles: produced from the nucleus in special laboratory conditions	$^1_1H = {}^1_1p$
	Neutron particles: produced in nuclear reactors	^1_0n
	Gamma rays: high-energy photons emitted from the nucleus of an atom	$^0_0\gamma$

Nuclear Reactions

In a **decay reaction**, a parent emits a particle or radiation to form the daughter nucleus. In the alpha decay reaction of radon-222, for example, the parent emits an alpha particle and a new daughter nucleus— polonium-218—is created.

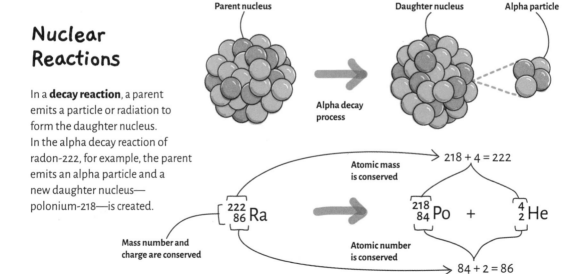

Parent nucleus   Daughter nucleus   Alpha particle

Alpha decay process

Atomic mass is conserved

$218 + 4 = 222$

$^{222}_{86}Ra$ $\longrightarrow$ $^{218}_{84}Po$ + ^4_2He

Mass number and charge are conserved

Atomic number is conserved

$84 + 2 = 86$

In a **capture reaction**, the nucleus captures a particle, which results in changes to the number of its nucleons.

A berillium-7 isotope captures one of its electrons closest to the nucleus. This negatively charged electron combines with one of the positively charged protons in the nucleus to form a neutron. The atomic number drops by one, creating the lithium-7 isotope.

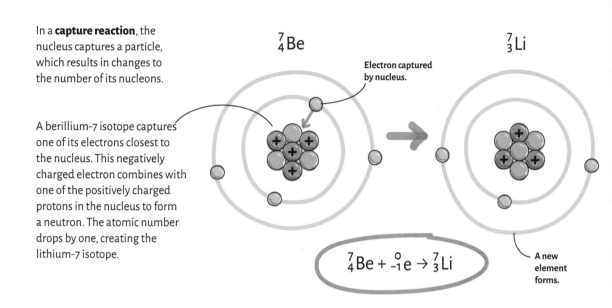

$$^7_4Be$$

$$^7_3Li$$

Electron captured by nucleus.

A new element forms.

$$^7_4Be + ^0_{-1}e \rightarrow ^7_3Li$$

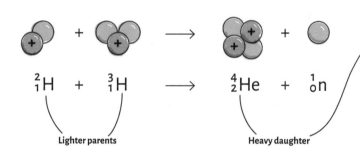

$$^2_1H + ^3_1H \longrightarrow ^4_2He + ^1_0n$$

Lighter parents

Heavy daughter

In a **fusion reaction**, two lighter parents combine to give birth to a heavy daughter and a lot of energy. When hydrogen-2 and hydrogen-3 isotopes undergo fusion, a heavier nucleus—helium-4—is created.

In a **fission reaction**, a heavier parent, such as uranium-235, splits into two or more lighter daughters. This releases an enormous amount of energy.

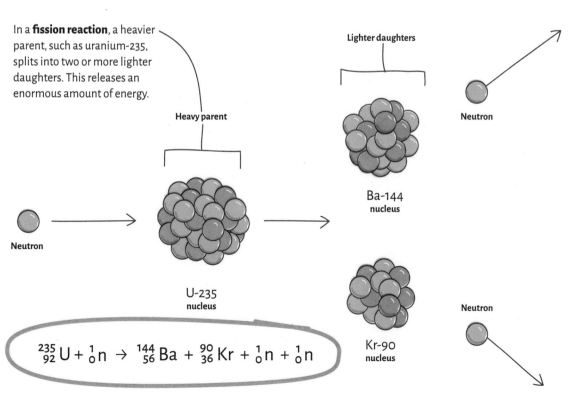

Lighter daughters

Heavy parent

Neutron

Ba-144
nucleus

Neutron

Neutron

U-235
nucleus

Kr-90
nucleus

$$^{235}_{92}U + ^1_0n \rightarrow ^{144}_{56}Ba + ^{90}_{36}Kr + ^1_0n + ^1_0n$$

HALF-LIFE AND USES OF RADIOISOTOPES

Radioisotopes contain unstable nuclei that undergo radioactive decay to form more stable atoms. Different radioisotopes lose their radioactivity at different rates, some more quickly than others. The degree of radioactivity or nuclear stability is quantified by **half-life**, which is the time required for half of the radioactive atoms in a sample to decay to their daughter isotopes.

Half-Life

The spontaneous decay of a naturally occurring isotope into a more stable form takes place in fixed time intervals during which a certain proportion of a radioactive sample disintegrates. This time interval is quantified by half-life, defined as the amount of time that must elapse for exactly half of a sample to decay. Iodine-131, for example, decays into xenon-131 by emitting an electron in a nuclear decay reaction with a half-life of eight days.

Suppose we start with 20 grams of the I-131 sample. After eight days, which is one half-life, 10 grams of the original I-131 sample remain as the other half turned into Xe-131. Each half-life that elapses takes away half of the remaining I-131 sample.

Iodine-131 is used to treat thyroid cancer. Its short half-life makes I-131 very suitable for medical applications because it does not remain in the human body for extended periods of times.

$$^{131}_{53} I \rightarrow {}^{131}_{54} Xe + {}^{0}_{-1} e$$

8 days   8 days

20 mg I-131   10 mg I-131   5 mg I-131

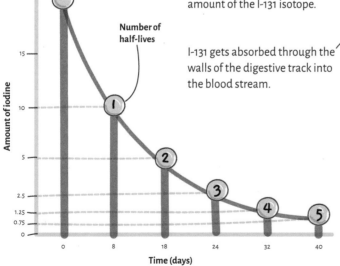

The patient ingests a predetermined amount of the I-131 isotope.

I-131 gets absorbed through the walls of the digestive track into the blood stream.

Number of half-lives

Thyroid gland

The thyroid gland produces the body's supply of iodine; it can easily accumulate the ingested I-131 isotope.

The I-131 isotope then decays by emitting radiation, which kills the surrounding cancerous cells with minimal damage to healthy cells.

Amount of iodine — 20, 15, 10, 5, 2.5, 1.25, 0.75, 0

Time (days) — 0, 8, 18, 24, 32, 40

Some radioisotopes can have half of their unstable nuclei decay in less than one second, while others can take many years. For example, the half-life of krypton-101 is one ten millionth of a second, whereas uranium-238 has a half-life of 4.51 billion years.

The half-life of a radioisotope is a primary factor in determining its potential technological applications.

RADIOISOTOPE	HALF-LIFE	APPLICATIONS
Technetium-99	6 hours	Brain, liver, lung, kidney imaging
Iron-59	45 days	Detection of anemia
Iodine-131	8 days	Thyroid therapy
Cobalt-60	5.3 years	Radiation cancer therapy
Uranium-235	7.04×10^8 years	Nuclear reactors
Carbon-14	5730 years	Archeological dating
Cesium-137	30 years	Cancer therapy

Nuclear Medicine

Radioisotopes with short half-lives are used for medical diagnosis and treatment. Relatively low doses of radioisotopes, called **tracers**, are needed for diagnostic medical imaging, but much stronger external radiation sources are utilized when treating cancerous tissue in **radiotherapy** applications.

Positron emission tomography (**PET**) utilizes fluorine-18 as a tracer to diagnose brain cancer. The tracer is injected into the patient and accumulates in the brain where it emits positrons— the antiparticle or the antimatter counterpart of an electron—that combine with electrons from the brain tissue releasing gamma rays.

Emitted gamma rays are detected and are used to generate an image that reveals normal and cancerous areas based on tissue activity.

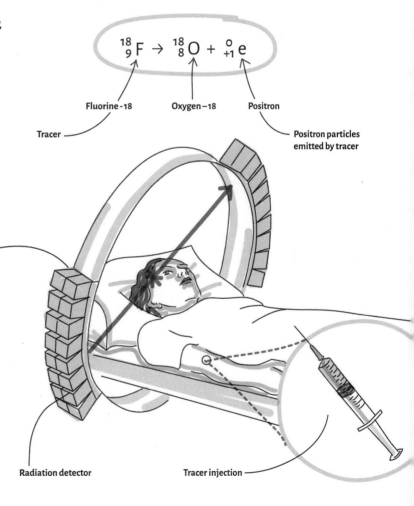

$$^{18}_{9}F \rightarrow ^{18}_{8}O + ^{0}_{+1}e$$

Fluorine-18   Oxygen-18   Positron

Tracer

Positron particles emitted by tracer

Radiation detector

Tracer injection

Isotopic Dating

As we know the half-life of many radioisotopes that are present in nature, dating archeological artifacts is a pretty straightforward technique, with radioactivity used to determine the age of rocks, minerals, plants, and various other fossils.

Carbon-14 dating is a common practice that scientists use to find the age of biological samples. Neutrons from cosmic rays collide with nitrogen-14 isotopes in the atmosphere to create carbon-14, which is consumed by plants in the form of carbon dioxide.

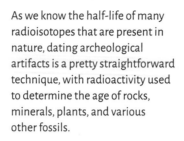

Cosmic radiations

Neutron

Animals and humans absorb carbon-14 by consuming plants. After death, living things can't accumulate carbon-14 any more, and the C-14 that is already present in the dead tissue starts to decay back to nitrogen-14. Examining the carbon-to-nitrogen ratio in a biological fossil therefore allows the age of the sample to be determined.

Neutron capture

Nitrogen-14

Carbon-14

Proton

Plants absorb carbon dioxide and incorporate carbon-14 through photosynthesis.

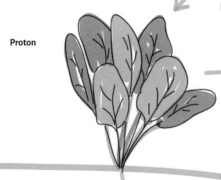

Animals and people eat plants and take in carbon-14.

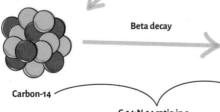

Beta decay

Carbon-14

Nitrogen-14

C-14:N-14 ratio in a sample provides age.

Nuclear Power

Nuclear power plants use the energy produced by the nuclear fission reaction of uranium-235 isotope to heat water into steam. The steam powers a large turbine, turning the heat energy into mechanical energy, which is converted into electrical energy by a generator.

The newly created neutrons hit additional uranium-235 isotopes in the reactor, yielding more neutrons and smaller nuclei. This **chain reaction** grows at an exponential rate and generates continuous energy.

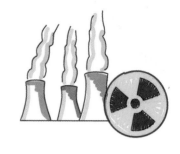

Uranium-235 is bombarded by neutrons to initiate a fission reaction in which two or three neutrons and smaller daughter nuclei are produced.

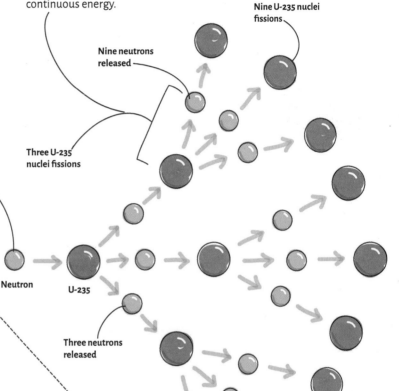

Nine U-235 nuclei fissions

Nine neutrons released

Three U-235 nuclei fissions

Accelerated neutron production in chain reaction

Neutron

U-235

Three neutrons released

Control rods determine the number of neutrons.

Steam released to power the turbine.

Steam generator

Fuel rods include U-235

Pressurized water circuit

Neutron absorbing rods are used in the nuclear reactor to make sure the fission reaction remains under control. Instead of generating huge amounts of energy, as in a nuclear bomb, smaller and steadier amounts of energy are produced.

Energy that keeps the nucleus together.

NUCLEAR BINDING ENERGY

NUCLEAR FORCE

A strong attractive force between nucleons.

THE NUCLEUS

NUCLEAR STABILITY

The balance between nuclear and electrostatic forces in the nucleus.

MASS DEFECT

Atomic mass that is converted to energy in nuclear reactions.

NUCLEAR CHEMISTRY

HALF-LIFE

Time required for radioactivity to drop by half.

NUCLEAR POWER

Generation of electricity from nuclear reactions.

HALF-LIFE AND USES OF RADIOISOTOPES

NUCLEAR MEDICINE

Use of radioisotopes for diagnosis and treatment.

ISOTOPIC DATING

Use of radioactivity to determine the age of fossils.

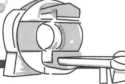

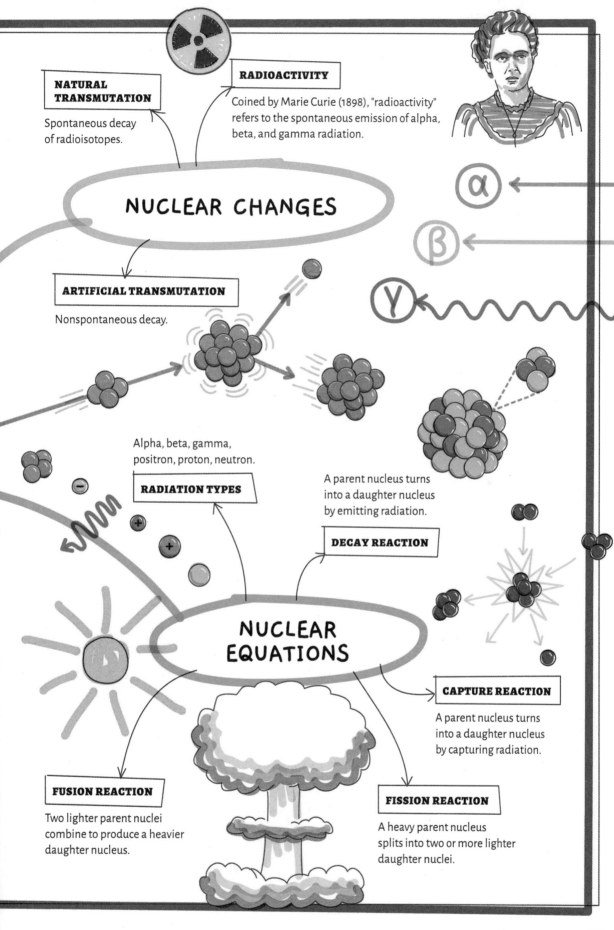

NATURAL TRANSMUTATION

Spontaneous decay of radioisotopes.

RADIOACTIVITY

Coined by Marie Curie (1898), "radioactivity" refers to the spontaneous emission of alpha, beta, and gamma radiation.

NUCLEAR CHANGES

ARTIFICIAL TRANSMUTATION

Nonspontaneous decay.

Alpha, beta, gamma, positron, proton, neutron.

RADIATION TYPES

A parent nucleus turns into a daughter nucleus by emitting radiation.

DECAY REACTION

NUCLEAR EQUATIONS

CAPTURE REACTION

A parent nucleus turns into a daughter nucleus by capturing radiation.

FUSION REACTION

Two lighter parent nuclei combine to produce a heavier daughter nucleus.

FISSION REACTION

A heavy parent nucleus splits into two or more lighter daughter nuclei.

CHAPTER 4

ELECTRONS IN ATOMS

Understanding electron behavior in atoms is key to explaining chemical bonding. Electrons are arranged in an ordered fashion outside of an atom's nucleus and not randomly distributed. They are particles with identical negative charge, mass, and volume but differ in their energy based on the proximity to the nucleus. Electrons with lowest energy are found closest to the nucleus where their attraction to the positively charged protons is the strongest. Electrons move around the nucleus as particles but also exhibit wave-like properties. This **particle-wave nature** of electrons is the key to understanding chemical bonding and reactivity.

ELECTROMAGNETIC RADIATION: LIGHT

Electromagnetic radiation is energy that travels through space as waves of massless "particles" called **photons**. The heat from a burning fire, the light from the sun, the X-rays used by your doctor, and the energy used to cook food in a microwave are all forms of electromagnetic radiation. Photons of different types of electromagnetic energy carry different energies. Unaided, the human eye can see only a tiny portion of the various types of electromagnetic radiation—we call this **visible light** or the **visible spectrum**.

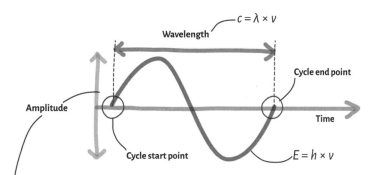

Photons in electromagnetic radiation travel through space in a wave form at a constant speed. This wave can be expressed in terms of **energy** (E), **wavelength** (λ), and **frequency** (v), which is measured in cycles per second using a unit of measurement known as a **Hertz** (Hz). Each of these three quantities are mathematically related to one another.

The **amplitude** (height) of the wave is associated with the brightness or intensity of the light: the higher the amplitude, the brighter the light. As the number of cycles traveling every second increases, the frequency increases. High-frequency light carries more energy than low-frequency light. As the wavelength increases, the frequency—and light energy—decreases.

A narrow range of light from infrared to ultraviolet (including visible light) is safe for humans in appropriate amounts, as the energy is low enough not to be harmful.

Beyond ultraviolet light, the high-frequency energy carried by photons is high enough to cause tissue damage. Light in this region is called **ionizing radiation**, as it can remove electrons from atoms and molecules in biological tissue. Light below the infrared region is nonionizing, as it does not carry enough energy to cause damage to the tissue.

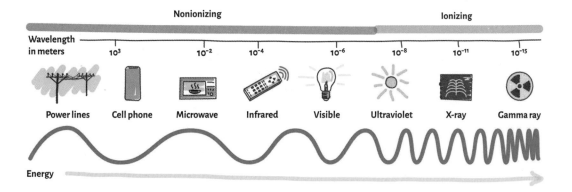

ELECTRON ARRANGEMENT: THE BOHR MODEL

During the 18th and 19th centuries, chemists performed flame tests in which they burned various compounds and used the color of the emitted light to identify elements. Following the discovery of the atomic structure, scientists were able to explain why each element was producing a different characteristic color. As the heat energy in these tests is not sufficient to cause changes in the nucleus of an atom, it meant that the observed colors must be a consequence of electron behavior outside of the nucleus.

In 1913, Niels Bohr proposed his description of electrons in an atom, which laid the groundwork for the quantum model. He explained that electrons are only allowed certain amounts of energy, called "quanta," and that they are located in orbits at discrete distances from the nucleus. Although Bohr's model only worked for the hydrogen atom, it still provided critical information for the more accurate models that followed.

Continuous and Line Spectra

When white light from an incandescent lamp is passed through a prism, a rainbow of colors can be seen from red to violet. Because each color diffuses into the next, from red to violet with no gaps in between, this observed color band is called a **continuous spectrum**.

When light emitted in a flame test or from a light bulb filled with a gaseous element is passed through a prism, something quite different is observed. Instead of a continuous spectrum, several bright and distinct lines of color are seen, each separated by darkness. This is called a **line spectrum**, in which some colors are certainly missing.

Each element in a flame test gives off a different observed color, and therefore each element will have a unique line spectrum that can be used for identification purposes.

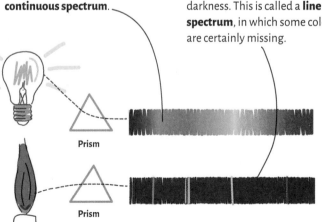

Prism

Prism

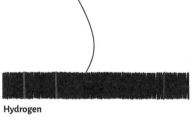

Hydrogen

Helium

Carbon

Bohr's Model

Bohr proposed his atomic model as a way of explaining the line spectra for various elements. According to the Bohr model, electrons are in circular orbits at distinct distances from the nucleus. These orbits are numbered by the **principal quantum number**, *n*. The first orbit closest to the nucleus is $n = 1$, which is lowest in energy. Energy increases as the distance from the nucleus—and the value of *n*—increases.

Electrons prefer to be closest to the nucleus, because of their electrostatic attraction to the positively charged nucleus. When electrons occupy the lowest energy orbits, the atom is in the **ground state**.

Electrons are unstable in the excited state. As an electron returns to the ground state, a photon of specific wavelength, frequency, and energy is emitted.

Heat in a flame test, or electricity in the case of a light bulb, energizes an electron in the ground state, forcing it to move up to higher energy orbits. Which higher energy orbit it moves to depends on how much external energy it absorbs, but the atom is now in an **excited state**.

Electron in an excited state.

Electron absorbing photon goes up energy level.

Increasing energy

$n = 3$

$n = 2$

$n = 1$

Nucleus

$\Delta E = E_3 - E_1$

Electron emitting photon goes down in energy level.

$$E_n = -2.178 \times 10^{-18}\,\text{J}\left(\frac{1}{n^2}\right)$$

Violet   Blue-green   Red

Large transition means higher energy.

$n = 5$

$n = 4$

$n = 3$

$n = 2$

$n = 1$

Small transition means lower energy.

The nature of the emitted electromagnetic radiation, and therefore the observed color, depends on how many orbits the electron jumps through during transition from the excited state to the ground state. The bigger the jump, the higher the energy of the emitted photon.

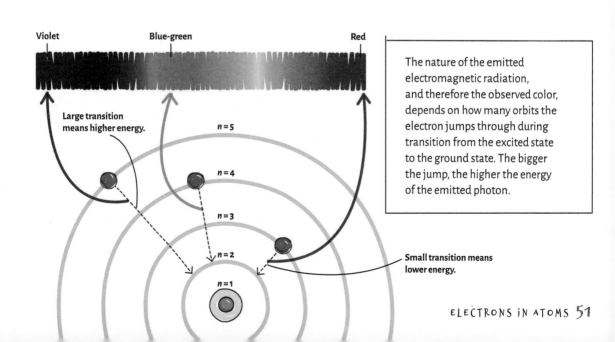

Electron Shell Distribution

The number of electrons that can occupy each orbit increases with the value of the principal quantum number. Bohr described the maximum number of electrons that can occupy each shell with the equation $2n^2$.

$n = 3 \rightarrow 2n^2 = 18$ electrons

$n = 2 \rightarrow 2n^2 = 8$ electrons

The number of electrons is the same as the atomic number, which is the number of protons in the nucleus. All of the electrons in an atom are distributed into energy orbits in the Bohr model.

$n = 1 \rightarrow 2n^2 = 2$ electrons

Nucleus

The principal quantum number, **n**, represents the period number (horizontal rows) of the periodic table. Lithium and fluorine both have electrons up to the second orbit, so they are both located in the second period.

Aluminum has electrons up to the third orbit, so it is located in the third period.

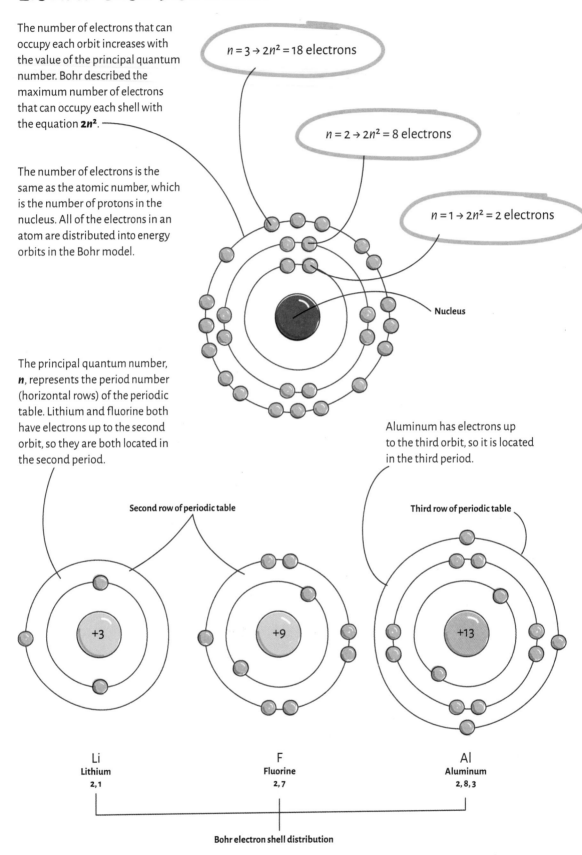

Second row of periodic table

Third row of periodic table

+3

+9

+13

Li
Lithium
2, 1

F
Fluorine
2, 7

Al
Aluminum
2, 8, 3

Bohr electron shell distribution

ELECTRON ARRANGEMENT: THE QUANTUM MODEL

Bohr's atomic model explained the line spectrum of hydrogen with one electron quite well, but was incapable of explaining the more complex spectra of multi-electron elements. The **quantum model** is a more sophisticated and accurate model that was developed later. In this model, the atom is essentially a probability map, where electrons are treated as both particles and waves. Instead of circling orbits, electrons are located in **orbitals** and their location is probabilistic instead of exact. Atoms have different numbers and shapes of orbitals depending upon their number of electrons, with the energy, shape, and other properties of these orbitals represented by **quantum numbers**.

Orbitals

Orbitals are three dimensional probability regions surrounding the nucleus, which show areas where an electron is likely to be found. Four orbital types (**s**, **p**, **d**, and **f**) are represented in the periodic table, but which of these are occupied by electrons in an atom depends on the number of electrons.

If we had a camera that was capable of photographing an electron as it zips around the nucleus, we could develop a picture that shows the locations where the electron had been. The shape of an orbital emerges when 90 percent of these locations are considered.

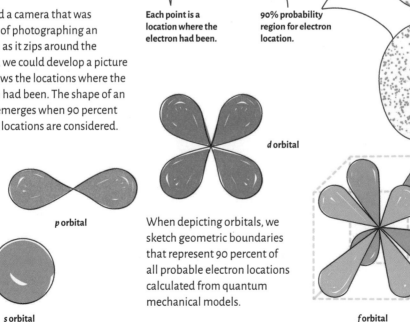

s orbital

Nucleus

p orbital

Each point is a location where the electron had been.

90% probability region for electron location.

Nucleus

d orbital

p orbital

s orbital

When depicting orbitals, we sketch geometric boundaries that represent 90 percent of all probable electron locations calculated from quantum mechanical models.

f orbital

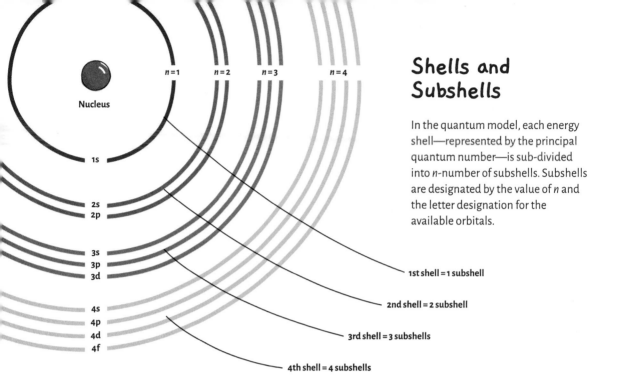

Nucleus

$n=1$ $n=2$ $n=3$ $n=4$

1s

2s
2p

3s
3p
3d

4s
4p
4d
4f

1st shell = 1 subshell

2nd shell = 2 subshell

3rd shell = 3 subshells

4th shell = 4 subshells

Shells and Subshells

In the quantum model, each energy shell—represented by the principal quantum number—is sub-divided into n-number of subshells. Subshells are designated by the value of n and the letter designation for the available orbitals.

Quantum Numbers

Quantum numbers are a series of numbers that are used to describe electrons in atoms, as well as the specific orbitals they are located in. An electron has four quantum numbers to describe it in an atom.

The first three locate the electrons in shells, subshells, and orbitals around the nucleus, similar to the way in which x, y, z coordinates determine the location of an object in space.

Electrons in orbitals behave as if they are tiny charged spheres spinning around an axis. This spin gives rise to a tiny magnetic field and to the fourth quantum number: the **magnetic spin quantum number (m_s)**.

NAME	SYMBOL	ALLOWED VALUES	MEANING
Principal quantum number	n	1, 2, 3, 4, …	Energy and size of shells
Angular momentum quantum number	l	0, 1, 2, …, $n-1$	Energy of sublevels and shape of orbitals $l = 0$ is s subshell $l = 1$ is p subshell $l = 2$ is d subshell $l = 3$ is f subshell
Magnetic quantum number	m_l	$-l, …, 0, …, +l$	Orbital orientation
Spin quantum number	m_s	$+^1/_2, -^1/_2$	Electron spin

The magnetic spin quantum number has only two possible values, $+\frac{1}{2}$ or $-\frac{1}{2}$, which indicate different directions of electron spin.

Quantum numbers determine the number of subshells, orbitals, and electrons for each principal quantum number.

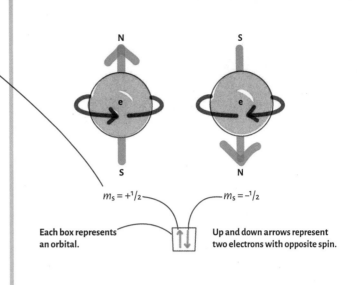

$m_s = +\frac{1}{2}$　　　$m_s = -\frac{1}{2}$

Each box represents an orbital.　　**Up and down arrows represent two electrons with opposite spin.**

Orbitals and electron spin

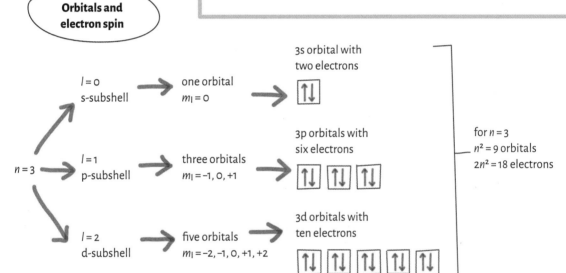

$n = 3$

$l = 0$ s-subshell → one orbital $m_l = 0$ → **3s orbital with two electrons**

$l = 1$ p-subshell → three orbitals $m_l = -1, 0, +1$ → **3p orbitals with six electrons**

$l = 2$ d-subshell → five orbitals $m_l = -2, -1, 0, +1, +2$ → **3d orbitals with ten electrons**

for $n = 3$
$n^2 = 9$ orbitals
$2n^2 = 18$ electrons

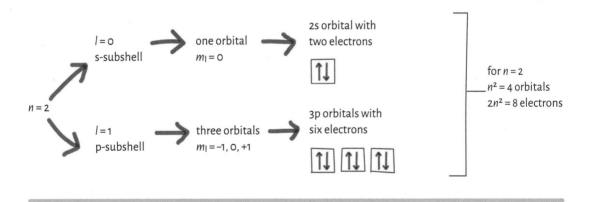

$n = 2$

$l = 0$ s-subshell → one orbital $m_l = 0$ → **2s orbital with two electrons**

$l = 1$ p-subshell → three orbitals $m_l = -1, 0, +1$ → **3p orbitals with six electrons**

for $n = 2$
$n^2 = 4$ orbitals
$2n^2 = 8$ electrons

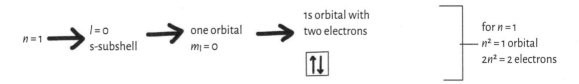

$n = 1$ → $l = 0$ s-subshell → one orbital $m_l = 0$ → **1s orbital with two electrons**

for $n = 1$
$n^2 = 1$ orbital
$2n^2 = 2$ electrons

ELECTRON CONFIGURATION

The distribution of electrons for an atom into orbitals is called its **electron configuration**. In 1925, Wolfgang Pauli discovered the principle that governs the arrangement of electrons in atoms according to the quantum model. Electrons in an atom tend to occupy the lowest energy levels available to them, due to their electrostatic attraction to the positively charged nucleus. The electron capacity for the Bohr orbits is followed, and electrons are distributed into orbitals when constructing electron configurations. This process is called the **Aufbau principle** (building up principle).

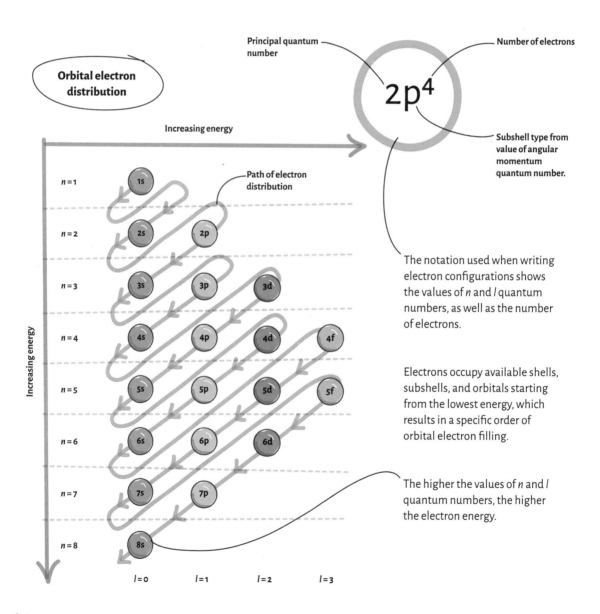

Orbital electron distribution

Principal quantum number

Number of electrons

$2p^4$

Subshell type from value of angular momentum quantum number.

Increasing energy

Path of electron distribution

The notation used when writing electron configurations shows the values of n and l quantum numbers, as well as the number of electrons.

Electrons occupy available shells, subshells, and orbitals starting from the lowest energy, which results in a specific order of orbital electron filling.

The higher the values of n and l quantum numbers, the higher the electron energy.

Increasing energy

$n = 1$ 1s

$n = 2$ 2s   2p

$n = 3$ 3s   3p   3d

$n = 4$ 4s   4p   4d   4f

$n = 5$ 5s   5p   5d   5f

$n = 6$ 6s   6p   6d

$n = 7$ 7s   7p

$n = 8$ 8s

$l = 0$ $l = 1$ $l = 2$ $l = 3$

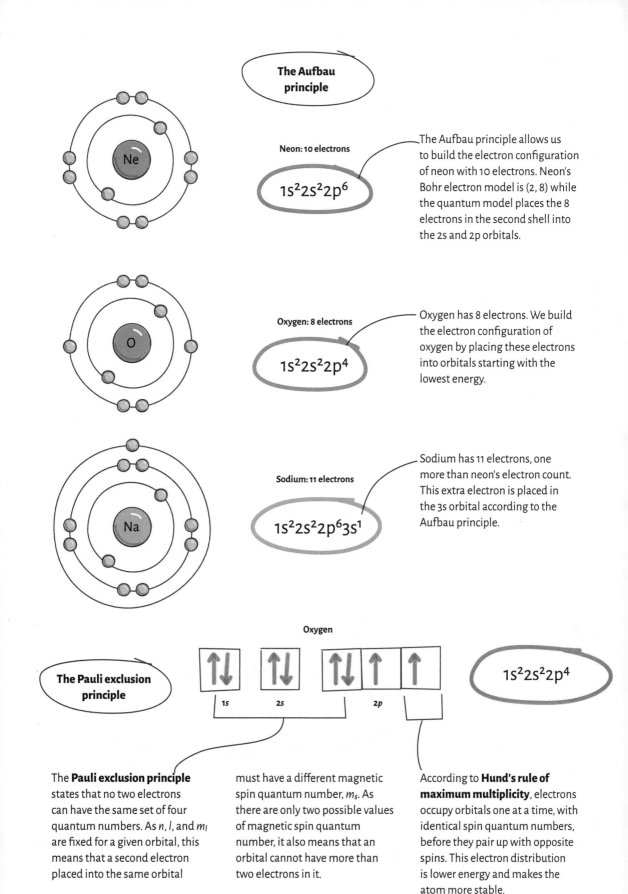

Neon: 10 electrons

$1s^2 2s^2 2p^6$

The Aufbau principle allows us
to build the electron configuration
of neon with 10 electrons. Neon's
Bohr electron model is (2, 8) while
the quantum model places the 8
electrons in the second shell into
the 2s and 2p orbitals.

Oxygen: 8 electrons

$1s^2 2s^2 2p^4$

Oxygen has 8 electrons. We build
the electron configuration of
oxygen by placing these electrons
into orbitals starting with the
lowest energy.

Sodium: 11 electrons

$1s^2 2s^2 2p^6 3s^1$

Sodium has 11 electrons, one
more than neon's electron count.
This extra electron is placed in
the 3s orbital according to the
Aufbau principle.

Oxygen

1s   2s   2p

**The Pauli exclusion
principle**

$1s^2 2s^2 2p^4$

The **Pauli exclusion principle**
states that no two electrons
can have the same set of four
quantum numbers. As n, l, and m_l
are fixed for a given orbital, this
means that a second electron
placed into the same orbital

must have a different magnetic
spin quantum number, m_s. As
there are only two possible values
of magnetic spin quantum
number, it also means that an
orbital cannot have more than
two electrons in it.

According to **Hund's rule of
maximum multiplicity**, electrons
occupy orbitals one at a time, with
identical spin quantum numbers,
before they pair up with opposite
spins. This electron distribution
is lower energy and makes the
atom more stable.

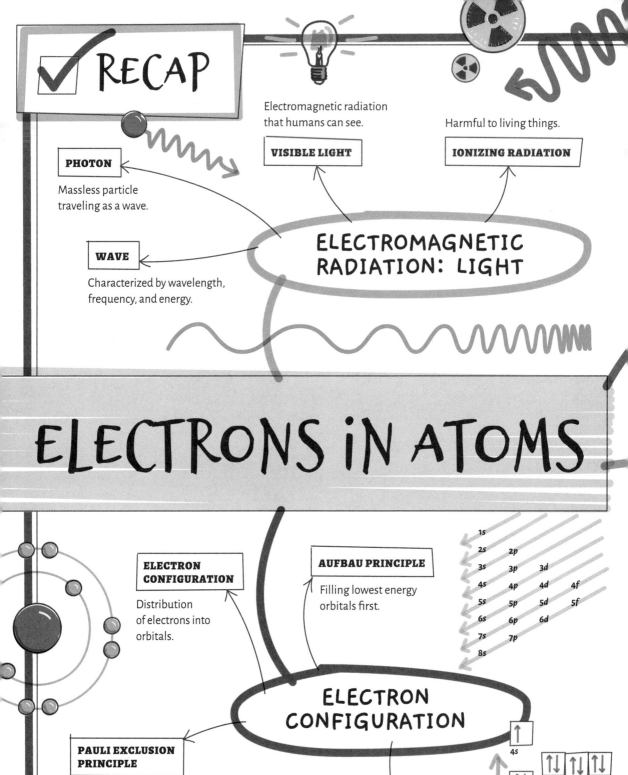

Electromagnetic radiation that humans can see.

VISIBLE LIGHT

Harmful to living things.

IONIZING RADIATION

PHOTON

Massless particle traveling as a wave.

WAVE

Characterized by wavelength, frequency, and energy.

ELECTROMAGNETIC RADIATION: LIGHT

ELECTRONS IN ATOMS

ELECTRON CONFIGURATION

Distribution of electrons into orbitals.

AUFBAU PRINCIPLE

Filling lowest energy orbitals first.

1s			
2s	2p		
3s	3p	3d	
4s	4p	4d	4f
5s	5p	5d	5f
6s	6p	6d	
7s	7p		
8s			

ELECTRON CONFIGURATION

PAULI EXCLUSION PRINCIPLE

No more than two electrons in an orbital.

HUND'S RULE OF MAXIMUM MULTIPLICITY

One electron in each orbital before pairing up.

Energy

4s ↑
3s ↑↓   2p ↑↓ ↑↓ ↑↓
2s ↑↓
1s ↑↓

↑↓ ↑↓ ↑↓ ↑ ↑

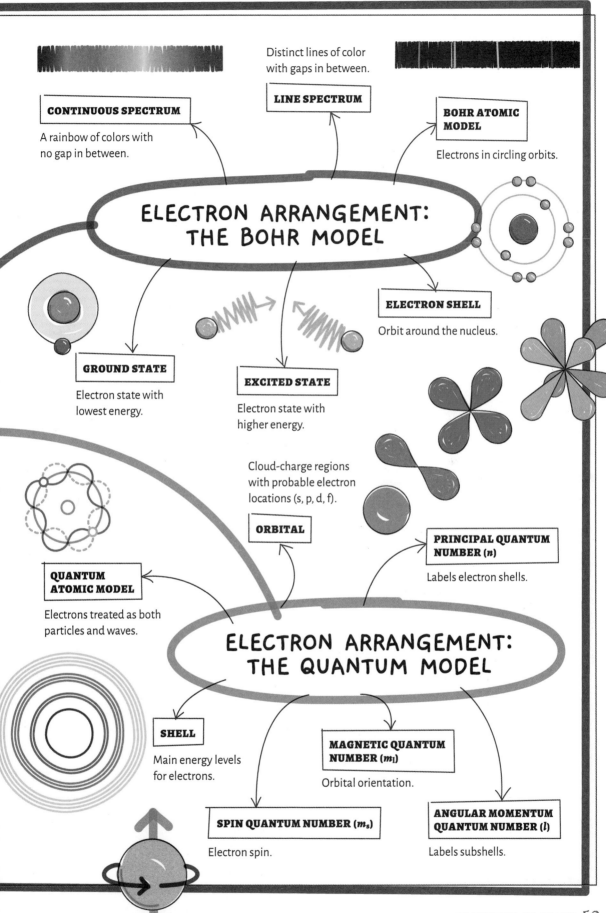

Distinct lines of color with gaps in between.

LINE SPECTRUM

CONTINUOUS SPECTRUM

A rainbow of colors with no gap in between.

BOHR ATOMIC MODEL

Electrons in circling orbits.

ELECTRON ARRANGEMENT: THE BOHR MODEL

ELECTRON SHELL

Orbit around the nucleus.

GROUND STATE

Electron state with lowest energy.

EXCITED STATE

Electron state with higher energy.

Cloud-charge regions with probable electron locations (s, p, d, f).

ORBITAL

PRINCIPAL QUANTUM NUMBER (n)

Labels electron shells.

QUANTUM ATOMIC MODEL

Electrons treated as both particles and waves.

ELECTRON ARRANGEMENT: THE QUANTUM MODEL

SHELL

Main energy levels for electrons.

MAGNETIC QUANTUM NUMBER (m_l)

Orbital orientation.

SPIN QUANTUM NUMBER (m_s)

Electron spin.

ANGULAR MOMENTUM QUANTUM NUMBER (l)

Labels subshells.

CHAPTER 5

THE PERIODIC TABLE OF THE ELEMENTS

The periodic table organizes elements based on their atomic number, so we can quickly look up the properties of an individual element, such as its mass, electron number, electron configuration, and unique chemical properties. The name "periodic table" comes from the fact that the table lists elements based on the periodic changes in their chemical properties. It arranges all known elements into **groups** and **periods**, which are closely related to their electron configuration: elements with similar chemical behavior are located in the same vertical group. Based on the elemental arrangement in the periodic table, clear trends in general properties emerge, making it quite convenient for scientists to efficiently discern the expected chemical nature of all elements.

QUANTUM NUMBERS AND THE PERIODIC TABLE

An element's location in the periodic table reflects the quantum numbers of its last orbital filled with electrons. The last shell occupied by electrons (the highest value of the principal quantum number or period number) is the **valence shell**, which is closely related to the chemical bonding properties. Elements in the periodic table belong to one of four blocks: **s-, p-, d-, or f-block**, based on their electron configuration. Elements located in the s- and p-blocks are classified as **main group elements**, as their electron configuration and chemical properties are predictable; **transition metals** are in the d-block; and **inner transition metals** are in the f-block.

The period where an element is located reflects the principal quantum number (n).

The block that an element belongs to provides the value of the angular momentum quantum number (l).

There are 2, 6, 10, and 14 groups in the **s-, p-, d-, and f-blocks**, respectively, for each period. These numbers reflect the electron capacity of the s, p, d, and f orbitals. The magnetic and spin quantum numbers (m_l and m_s) are represented by each box within a block.

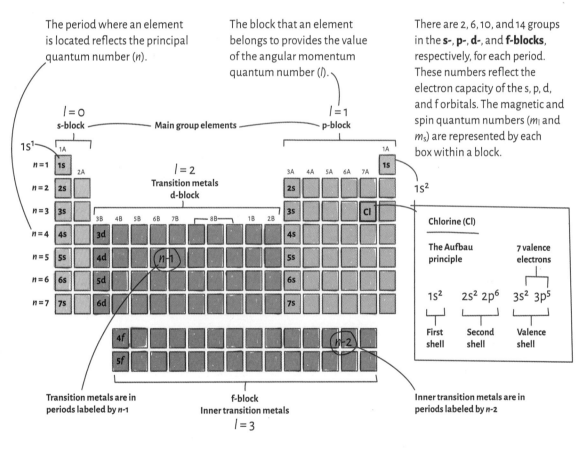

Transition metals are in periods labeled by n-1

f-block
Inner transition metals
$l = 3$

Inner transition metals are in periods labeled by n-2

The Aufbau principle is used to fill all orbitals with electrons until the desired number of electrons is reached. This provides the location of the element in the periodic table.

The atomic number of chlorine (Cl) is 17, with 17 electrons. It is located in the third period (n=3), which is its valence shell reflected in its electron configuration.

Chlorine has seven electrons in the valence shell (**valence electrons**), which places it in group 7A of the periodic table—the seventh vertical main group.

STABLE ELECTRON CONFIGURATIONS

During the 19th century, chemists organized all known elements according to how they chemically bonded with other elements. Their observations showed that a specific group of elements known as the **noble gases** (located in group 8A of the periodic table) were found in nature in their pure elemental form. This means they exhibit little or no reactivity with other elements; they are chemically quite *inert* or *stable*. Since chemical reactivity is related directly to electron configuration, noble gases must therefore have a very stable electron configuration. It was later discovered that all noble gases—with the exception of helium—have eight valence electrons in their outermost shell (helium has two valence electrons).

Electron Configurations from the Periodic Table

Since noble gases are stable and chemically inert, all other elements try to follow their electron configuration. They do this by holding either two valence electrons in their outermost shell, like helium (the **duplet rule**), or eight valence electrons, like the rest of the noble gases (the **octet rule**).

Each period closes with a noble gas whose available orbitals are all filled completely. This **closed-shell** electron configuration for noble gases is shown by writing the elemental symbol in brackets.

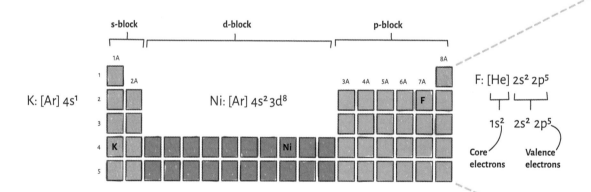

Electron configurations can easily be written based on the location of an element in the periodic table.

For fluorine (F), **full electron configuration** is written by filling each orbital starting with the 1s orbital (the first box for hydrogen) until fluorine is reached on the periodic table with nine electrons.

Fluorine is in the second period; the first period closes with helium. The **abbreviated electron configuration** of fluorine is written by expressing the first period with [He] representing the core electrons.

A **stable electron configuration** refers to an atom or ion in which the valence shell is completely filled with electrons. Noble gases have a stable electron configuration. All other elements tend to gain or lose electrons in order to attain a stable electron configuration.

A sodium atom loses one of its 11 electrons to become a **cation** with the same electron configuration as neon (10 electrons).

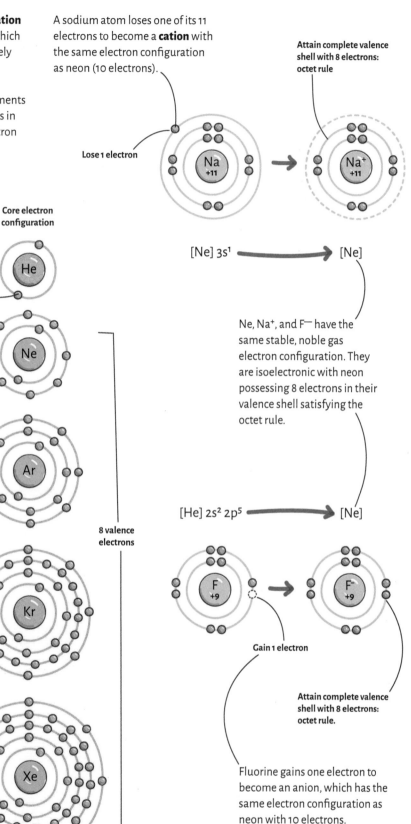

Attain complete valence shell with 8 electrons: octet rule

Lose 1 electron

Na +11

Na⁺ +11

Core electron configuration

2 valence electrons

He

$[Ne]\ 3s^1$ $[Ne]$

Ne, Na⁺, and F⁻ have the same stable, noble gas electron configuration. They are isoelectronic with neon possessing 8 electrons in their valence shell satisfying the octet rule.

Ne

8 valence electrons

Ar

$[He]\ 2s^2\ 2p^5$ $[Ne]$

2	
He	

Kr

F +9

F⁻ +9

10	
Ne	

Gain 1 electron

Attain complete valence shell with 8 electrons: octet rule.

18	
Ar	

36	
Kr	

Xe

54	
Xe	

Fluorine gains one electron to become an anion, which has the same electron configuration as neon with 10 electrons.

For main group elements, the group number equals the number of valence electrons. This makes it easy to predict how many electrons will be lost or gained to obtain the stable octet of 8 valence electrons.

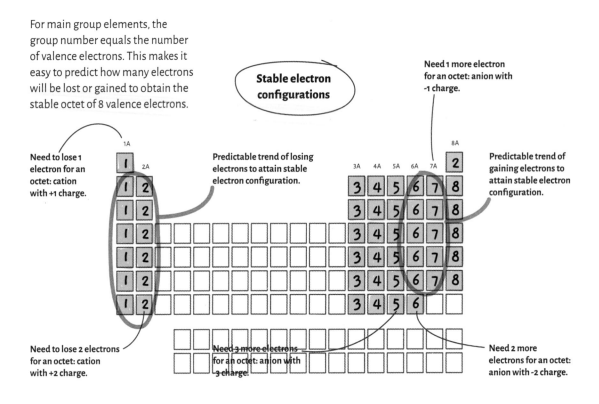

Need to lose 1 electron for an octet: cation with +1 charge.

Stable electron configurations

Need 1 more electron for an octet: anion with -1 charge.

Predictable trend of losing electrons to attain stable electron configuration.

Predictable trend of gaining electrons to attain stable electron configuration.

Need to lose 2 electrons for an octet: cation with +2 charge.

Need 3 more electrons for an octet: anion with -3 charge.

Need 2 more electrons for an octet: anion with -2 charge.

Writing Electron Configuration for Ions

Atoms gain or lose electrons to form ions to become more stable by obtaining a stable electron configuration like the noble gases. When writing ionic electron configurations, electrons are added to or removed from the highest-energy valence shell. Therefore, only valence electrons are involved in the formation of ions.

Electrons are **removed** from the valence shell when writing the electron configuration of a **cation**.

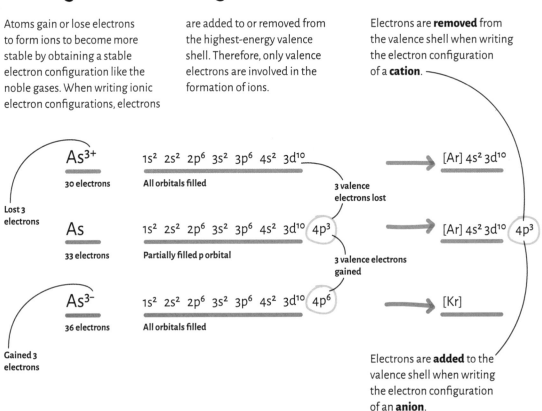

As^{3+}

30 electrons

All orbitals filled

$1s^2\ 2s^2\ 2p^6\ 3s^2\ 3p^6\ 4s^2\ 3d^{10}$

3 valence electrons lost

$[Ar]\ 4s^2\ 3d^{10}$

Lost 3 electrons

As

33 electrons

Partially filled p orbital

$1s^2\ 2s^2\ 2p^6\ 3s^2\ 3p^6\ 4s^2\ 3d^{10}\ 4p^3$

3 valence electrons gained

$[Ar]\ 4s^2\ 3d^{10}\ 4p^3$

As^{3-}

36 electrons

All orbitals filled

$1s^2\ 2s^2\ 2p^6\ 3s^2\ 3p^6\ 4s^2\ 3d^{10}\ 4p^6$

$[Kr]$

Gained 3 electrons

Electrons are **added** to the valence shell when writing the electron configuration of an **anion**.

PERIODIC CLASSIFICATION OF ELEMENTS

As valence electrons are responsible for observed chemical properties, elements with a similar chemical nature are located in the same group of the periodic table. Atoms tend to lose, gain, or share electrons in order to obtain a stable electron configuration. Therefore, elements that are in groups closest to the noble gases tend to be more reactive compared to those farther away. This periodic trend of varying chemical properties enables classification of all elements into **metals**, **nonmetals**, and **metalloids** (semi-metals).

METALS
- tend to form cations
- shiny luster
- malleable
- ductile
- good conductors of heat and electricity
- mostly solid at room temperature
- high density

METALLOIDS
- may gain or lose electrons
- properties in between metals and nonmetals
- some are shiny, some are dull
- some are malleable some are not
- some are ductile, some are not
- semiconductive to electrical charge

NONMETALS
- tend to form anions except for noble gases
- dull
- brittle
- not malleable
- not ductile
- poor conductors of heat and electricity
- can be solid, liquid, or gas at room temperature

The proximity of **alkali metals** in group 1A to the noble gases makes them the most reactive metals. They tend to be soft and shiny, and need special storage; they are never found in pure form in nature.

Alkaline earth metals are not as reactive as alkali metals, but still have more reactivity than most metals. They tend to be fairly hard and bright white, and are found in nature combined with other elements.

Halogens are highly reactive nonmetals due to their proximity to the noble gases. Compounds containing halogens are called "halides."

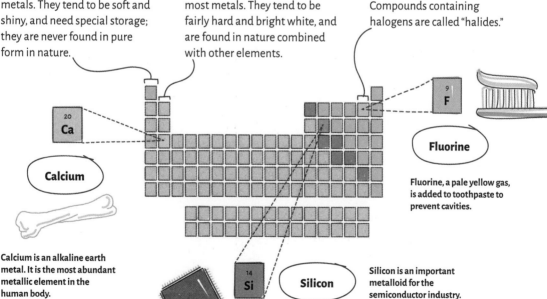

Calcium

Fluorine

Fluorine, a pale yellow gas, is added to toothpaste to prevent cavities.

Calcium is an alkaline earth metal. It is the most abundant metallic element in the human body.

Silicon

Silicon is an important metalloid for the semiconductor industry.

PERIODIC TRENDS

Periodic trends are predictable patterns that demonstrate different aspects of elements based on their location in the periodic table. **Atomic radius, ionization energy, electron affinity, electronegativity,** and **metallic character** are the most important trends; they provide chemists with an invaluable tool to quickly predict elemental properties. These predictable trends exist because of the similar atomic structure of elements distributed into groups and periods based on their electron configuration.

Atomic radius gets larger as you move down a group, as there are more filled or partially-filled electron shells.

Atoms get smaller as you move right across a period. This is because no new shells of electrons are added, and the increasing positive charge of the nucleus pulls electrons in closer to the nucleus, forcing the atom to shrink in size.

Ionization energy is the energy required to remove an electron from a neutral atom in its gaseous state. Valence electrons are farther away from the nucleus in larger atoms, and less energy is required to remove them. The smaller the atom, the higher the ionization energy.

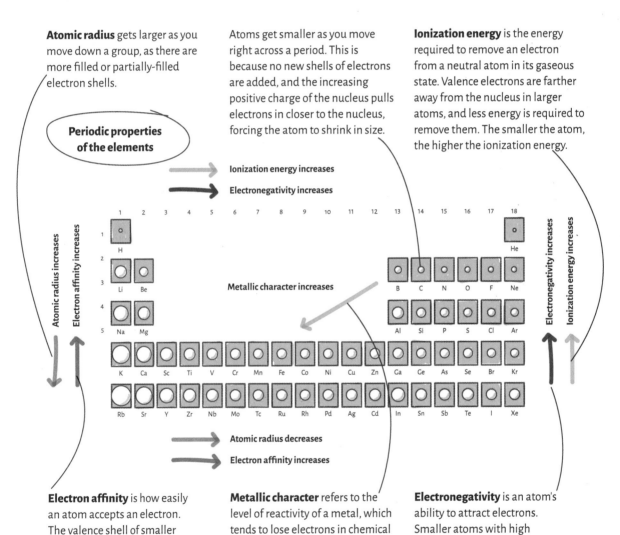

Electron affinity is how easily an atom accepts an electron. The valence shell of smaller atoms is closer and more strongly attracted to the nucleus, so it is easier to add electrons to them. The smaller the atom, the higher the electron affinity.

Metallic character refers to the level of reactivity of a metal, which tends to lose electrons in chemical reactions. Metallic character generally increases diagonally from the top right of the periodic table toward the bottom left.

Electronegativity is an atom's ability to attract electrons. Smaller atoms with high electron affinity have high electronegativity. Fluorine is the most electronegative element on the periodic table.

LEWIS DOT SYMBOLS

Valence electrons are involved in most chemical reactions in which chemical bonds between atoms are broken and formed. The American chemist G. N. Lewis devised a brilliant system of symbols (now known as **Lewis dot diagrams**) to represent the valence electrons and provide a better understanding of how an atom bonds to other atoms to form compounds. Lewis dot symbols easily explain how an atom tends to follow the octet rule to obtain a stable electron configuration, as well as how ions and compounds form for main group elements.

Writing Lewis Dot Symbols

In a **Lewis dot symbol**, an element's symbol represents the core of the atom (core electrons and the nucleus) and valence electrons are represented by one dot for each electron. There are four sides to each elemental symbol, and these can hold up to two valence electrons for a total of eight (octet rule).

For main group elements, the Lewis dot symbols conveniently show what an atom needs to do in order to obtain a stable, noble gas electron configuration.

STEP 1:
Locate element in the periodic table and write its symbol.

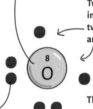

Oxygen 8 electrons.

STEP 2:
Determine the number of valence electrons. Oxygen is a main group element located in group 6A (or 16). It has 6 valence electrons

Each dot represents one valence electron up to 8 for octet.

Two available spots indicate oxygen needs two more electrons for an octet.

STEP 3:
Place a single dot for each valence electron to all four sides of the atomic symbol for oxygen before pairing up the dots for 6 valence electrons.

The element's symbol represents core electrons and the nucleus.

Beryllium (Be) and boron (B) are two odd members of the main group elements, as they do not follow the duplet or octet rule. These elements are perfectly satisfied with two and three valence electrons, respectively.

Elements located in groups closer to the noble gases tend to be more reactive. Groups 1A and 7A are highly reactive, as they are both one valence electron away from the noble gas configuration.

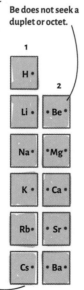

Be does not seek a duplet or octet.

B does not seek a duplet or octet.

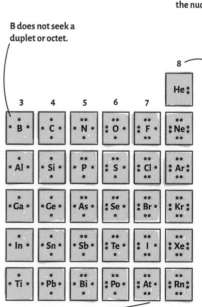

Number of valence electrons.

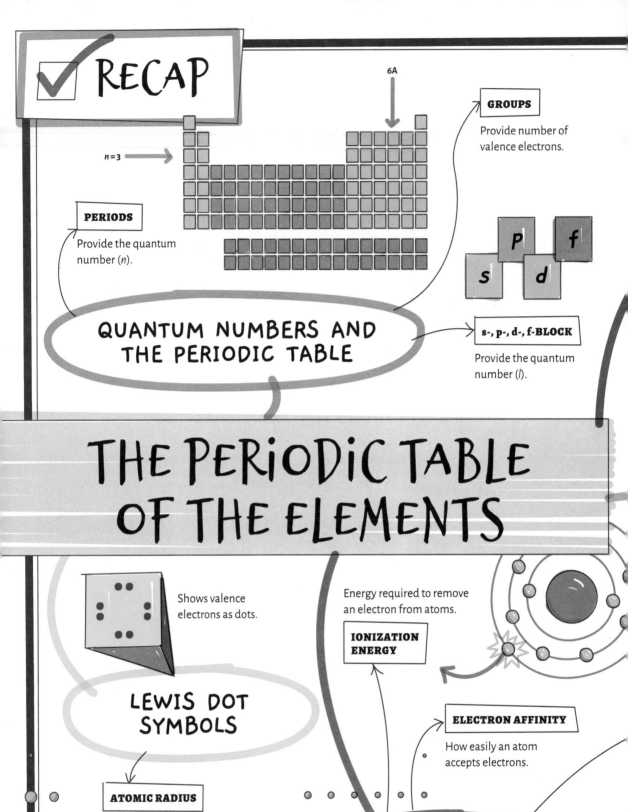

RECAP

6A

GROUPS

Provide number of valence electrons.

$n=3$

PERIODS

Provide the quantum number (n).

p f

s d

QUANTUM NUMBERS AND THE PERIODIC TABLE

s-, p-, d-, f-BLOCK

Provide the quantum number (l).

THE PERIODIC TABLE OF THE ELEMENTS

Shows valence electrons as dots.

Energy required to remove an electron from atoms.

IONIZATION ENERGY

LEWIS DOT SYMBOLS

ELECTRON AFFINITY

How easily an atom accepts electrons.

ATOMIC RADIUS

Increases down a group; decreases across a period from left to right.

PERIODIC TRENDS

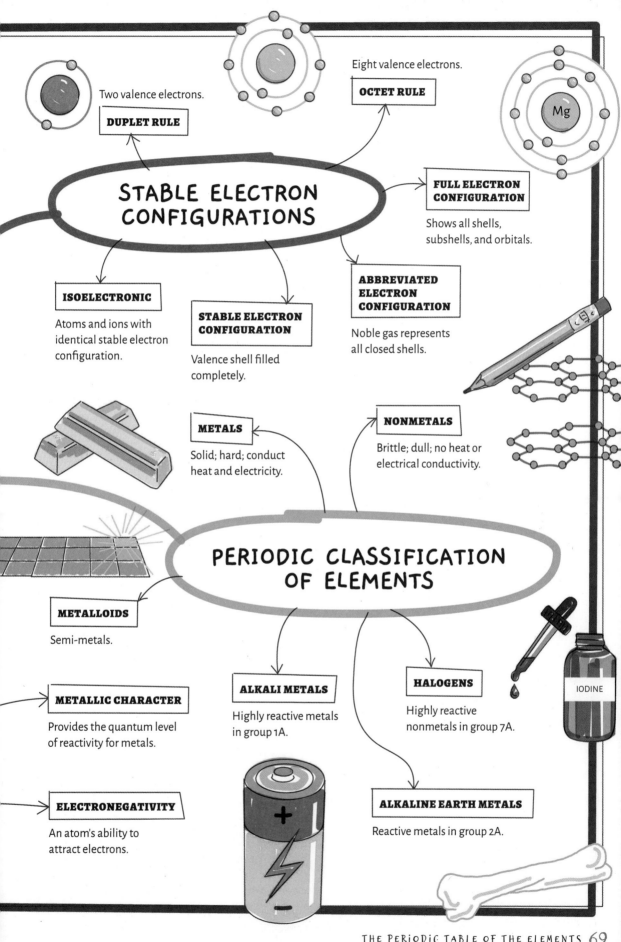

Two valence electrons.

DUPLET RULE

Eight valence electrons.

OCTET RULE

Mg

STABLE ELECTRON CONFIGURATIONS

FULL ELECTRON CONFIGURATION

Shows all shells, subshells, and orbitals.

ISOELECTRONIC

Atoms and ions with identical stable electron configuration.

STABLE ELECTRON CONFIGURATION

Valence shell filled completely.

ABBREVIATED ELECTRON CONFIGURATION

Noble gas represents all closed shells.

METALS

Solid; hard; conduct heat and electricity.

NONMETALS

Brittle; dull; no heat or electrical conductivity.

PERIODIC CLASSIFICATION OF ELEMENTS

METALLOIDS

Semi-metals.

METALLIC CHARACTER

Provides the quantum level of reactivity for metals.

ALKALI METALS

Highly reactive metals in group 1A.

HALOGENS

Highly reactive nonmetals in group 7A.

IODINE

ELECTRONEGATIVITY

An atom's ability to attract electrons.

ALKALINE EARTH METALS

Reactive metals in group 2A.

CHAPTER 6

CHEMICAL BONDING

A **chemical bond** is a lasting attraction between atoms, ions, or molecules that leads to the formation of chemical compounds. Chemical bonding—one of the most fundamental concepts of chemistry—is essential in explaining other important concepts, such as reactivity and the properties of matter. There are chemical bonding theories based on experimental observations that explain why atoms are attracted to each other and how products form in a chemical reaction. When atoms approach each other, their valence electrons interact and get redistributed; if the resulting energy of the bonded atoms is lower than the sum of the energies of the component atoms, a stable chemical bond is formed.

TYPES OF CHEMICAL BOND

There are 91 naturally occurring elements. It would be impossible to think that these could make up all matter and life in their pure form; instead, they must combine with each other via chemical bonding to create a vast number of other compounds. Chemical bonding can generally be divided into three fundamentally different types: **ionic**, **covalent**, and **metallic** bonding. The type of bonding depends on the nature of the atoms involved and largely determines the physical and chemical properties of matter.

Metallic bonding occurs between two metal atoms and involves delocalized, loosely bonded valence electrons.

Covalent bonding occurs between two nonmetal atoms, or between a nonmetal atom and a metalloid atom. It involves a *sharing of valence electrons*.

Ionic bonding occurs between a metal atom and a nonmetal atom. It involves a *complete transfer of valence electrons* from the metal atom to the nonmetal atom.

Electronegativity plays a crucial role in what type of bonding is favored between two atoms: an ionic bond is rarely purely ionic, and a covalent bond is rarely 100 percent covalent.

The nature and characteristics of a chemical bond change, depending on the difference in electronegativity of the atoms involved in the bonding.

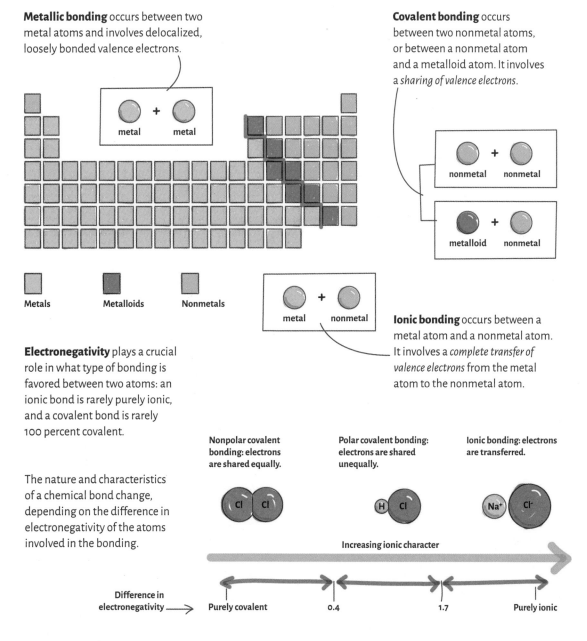

Metals   Metalloids   Nonmetals

metal + metal

nonmetal + nonmetal

metalloid + nonmetal

metal + nonmetal

Nonpolar covalent bonding: electrons are shared equally.

Polar covalent bonding: electrons are shared unequally.

Ionic bonding: electrons are transferred.

Cl — Cl

H — Cl

Na⁺ — Cl⁻

Increasing ionic character

Difference in electronegativity ⟶   Purely covalent   0.4   1.7   Purely ionic

iONiC BONDiNG AND iONiC COMPOUNDS

onic bonding occurs when a metal atom with lower electronegativity bonds with a nonmetal atom with higher electronegativity: the greater the difference in electronegativity, the stronger the ionic character of the bond that forms between the atoms. The metal atom loses valence electrons, becoming a cation, while the nonmetal atom accepts those electrons to form an anion. The strong attraction between the oppositely charged ions is called an ionic bond and the resulting compound is an ionic compound. The number of electrons that are lost and gained must be identical for a charge-neutral ionic compound to form.

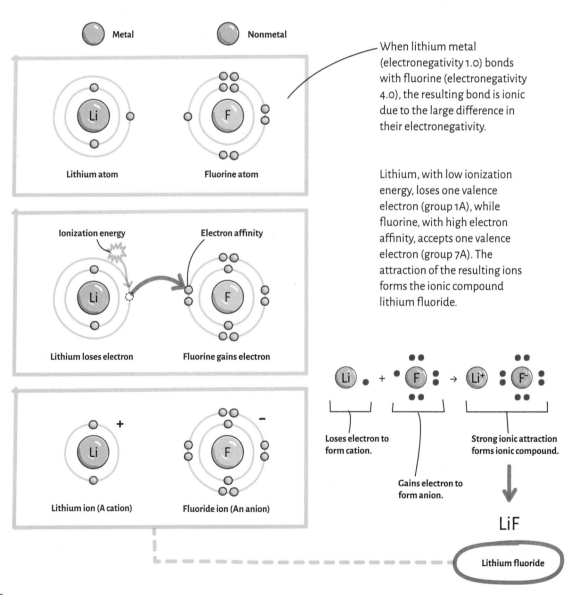

Metal

Nonmetal

Lithium atom

Fluorine atom

When lithium metal (electronegativity 1.0) bonds with fluorine (electronegativity 4.0), the resulting bond is ionic due to the large difference in their electronegativity.

Lithium, with low ionization energy, loses one valence electron (group 1A), while fluorine, with high electron affinity, accepts one valence electron (group 7A). The attraction of the resulting ions forms the ionic compound lithium fluoride.

Ionization energy

Electron affinity

Lithium loses electron

Fluorine gains electron

Lithium ion (A cation)

Fluoride ion (An anion)

Li • + • F : → Li⁺ : F⁻ :

Loses electron to form cation.

Gains electron to form anion.

Strong ionic attraction forms ionic compound.

LiF

Lithium fluoride

Binary Ionic Compounds

Binary ionic compounds form between two monoatomic ions of main group elements whose ionic charges are predictable.

Crossed over ionic charges appear as superscripts in the formula, where the metal is always written first followed by the nonmetal.

The charge of the cation of the metal determines how many anions should be present in the charge-neutral ionic compound.

The charge of the anion from the nonmetal determines how many cations should be present in the charge-neutral ionic compound.

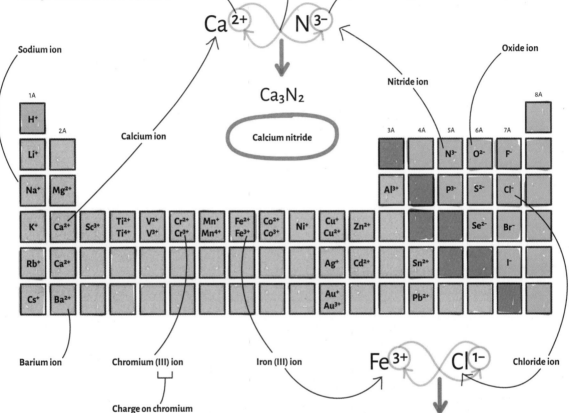

Sodium ion

Oxide ion

Nitride ion

Calcium ion

Calcium nitride

Barium ion

Chromium (III) ion

Charge on chromium

Iron (III) ion

Chloride ion

Iron (III) chloride

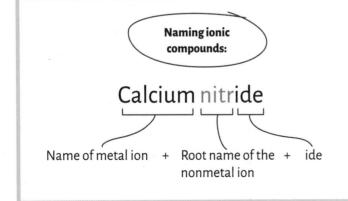

Naming ionic compounds:

Calcium nitride

Name of metal ion + Root name of the + ide
nonmetal ion

For transition metals, ionic charges are variable. Iron, for example, can have a +2 charge and a +3 charge in its ionic forms. The nature of the charge is specified in Roman numerals following the name of the transition metal, such as iron (II) and iron (III).

Ternary Ionic Compounds

Ternary ionic compounds form between at least three different elements. They may include a polyatomic ion and at least one metal or one nonmetal atom, or two different polyatomic ions.

Polyatomic ions are a group of two or more covalently-bonded atoms that carry a positive or negative charge.

Elements in polyatomic ions are covalently bonded to form a molecule with an overall positive or negative charge.

For example, the active ingredient in household bleach is sodium hypochlorite (NaClO). Here, the hypochlorite ion (ClO$^-$) is a polyatomic ion where the negative charge belongs to the entire molecule and not to oxygen alone.

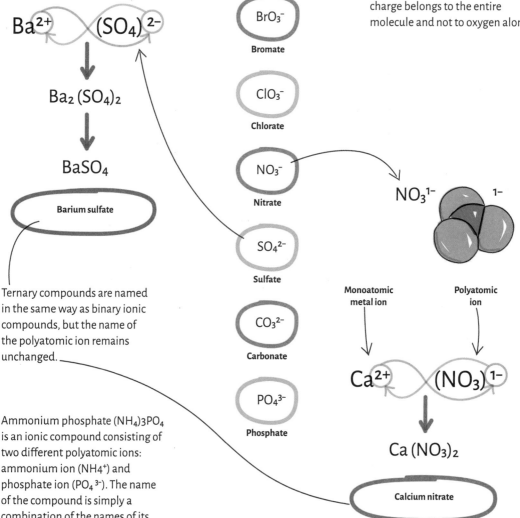

Common polyatomic ions

BrO$_3^-$
Bromate

ClO$_3^-$
Chlorate

NO$_3^-$
Nitrate

SO$_4^{2-}$
Sulfate

CO$_3^{2-}$
Carbonate

PO$_4^{3-}$
Phosphate

Ba^{2+} × $(SO_4)^{2-}$

$Ba_2(SO_4)_2$

$BaSO_4$

Barium sulfate

NO_3^{1-} $1-$

Monoatomic metal ion   Polyatomic ion

Ca^{2+} × $(NO_3)^{1-}$

$Ca(NO_3)_2$

Calcium nitrate

Ternary compounds are named in the same way as binary ionic compounds, but the name of the polyatomic ion remains unchanged.

Ammonium phosphate (NH$_4$)3PO$_4$ is an ionic compound consisting of two different polyatomic ions: ammonium ion (NH4$^+$) and phosphate ion (PO$_4$ $^{3-}$). The name of the compound is simply a combination of the names of its polyatomic ions with no changes.

Ammonium phosphate is an important ionic compound commonly used as an ingredient in certain types of fertilizers to supply elemental nitrogen necessary for plant growth.

When a transition metal is involved, the charge on the transition metal is provided in Roman numerals in ternary ionic compounds such as in copper (II) sulfate (CuSO$_4$).

Copper (II) sulfate is used in swimming pools to prevent algae growth and the spread of athlete's foot.

COVALENT BONDING AND MOLECULAR COMPOUNDS

Nonmetals typically have a higher electronegativity than the rest of the periodic table, and tend to have a high electron affinity due to their small atomic size. Consequently, when a nonmetal bonds with another nonmetal or metalloid, neither atom transfers electrons to the other: in order to seek a stable electron configuration, the two atoms share their valence electrons. The shared electrons interact with both of the nuclei of the bonding atoms, lowering their potential energy. This interaction is a **covalent bond**. A **covalent compound** or a **molecular compound** forms as a result of covalent bonding.

Formation of a Covalent Bond

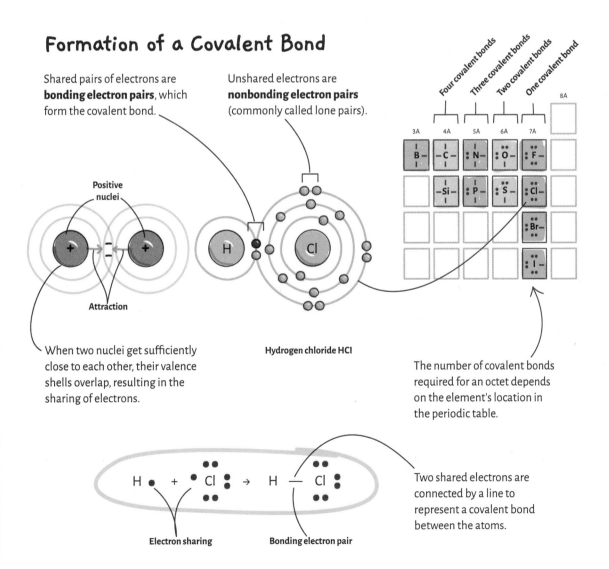

Shared pairs of electrons are **bonding electron pairs**, which form the covalent bond.

Unshared electrons are **nonbonding electron pairs** (commonly called lone pairs).

Positive nuclei

Attraction

When two nuclei get sufficiently close to each other, their valence shells overlap, resulting in the sharing of electrons.

Hydrogen chloride HCl

The number of covalent bonds required for an octet depends on the element's location in the periodic table.

Two shared electrons are connected by a line to represent a covalent bond between the atoms.

Electron sharing

Bonding electron pair

Nonmetals can share one, two, or three pairs of electrons, forming **single**, **double**, or **triple covalent bonds**, respectively, to satisfy the octet or duplet rule.

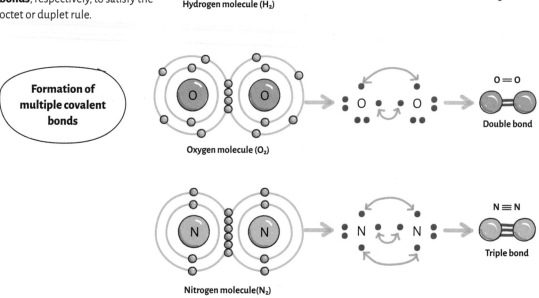

Formation of multiple covalent bonds

Hydrogen molecule (H_2)

Single bond

Oxygen molecule (O_2)

Double bond

Nitrogen molecule(N_2)

Triple bond

Polar and Nonpolar Covalent Bonds

If the electronegativity difference between the bonded atoms is less than 0.4, a **nonpolar** covalent bond forms, meaning electrons are shared equally by both atoms.

If the electronegativity difference between the bonded atoms is between 0.4 and 1.7, a **polar covalent bond** forms, meaning electrons are not shared equally.

The atom with higher electronegativity pulls the shared electrons to itself more strongly, creating a region of higher negative charge density.

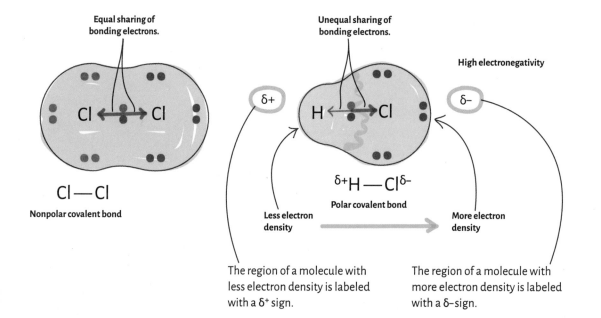

Equal sharing of bonding electrons.

Cl — Cl

Nonpolar covalent bond

Unequal sharing of bonding electrons.

High electronegativity

$\delta+$

$\delta-$

$\delta+$H — Cl$\delta-$

Polar covalent bond

Less electron density

More electron density

The region of a molecule with less electron density is labeled with a $\delta+$ sign.

The region of a molecule with more electron density is labeled with a $\delta-$ sign.

Molecular Compounds

In **molecular compounds** or **covalent compounds**, atoms are bonded together via covalent bonding.

The laws of definite and multiple proportions (see page 26) are upheld in molecular compounds.

PREFIXES

1	mono	6	hexa
2	di	7	hepta
3	tri	8	octa
4	tetra	9	nona
5	penta	10	deca

NO_2
Nitrogen dioxide

N_2O
Dinitrogen monoxide

When naming molecular compounds, prefixes are added to indicate the number of each element present in the molecule.

SiH_4
Silicon tetrahydride

CO_2
Carbon dioxide

If there is only one of the first atoms, there is no need to begin the name with "mono."

CO
Carbon monoxide

OF_2
Oxygen difluoride

Element farthest to right or top in periodic table.

Element furthest to left or bottom in periodic table.

I_2O_5
Diiodine pentoxide

The least electronegative element (fartherst toward the left and bottom of the periodic table) is always written first in molecular formulas of covalently bonded compounds.

The most electronegative element takes the -*ide* ending.

Naming molecular compounds:

Diiodine pentoxide

Prefix + Element name + Root name + ide

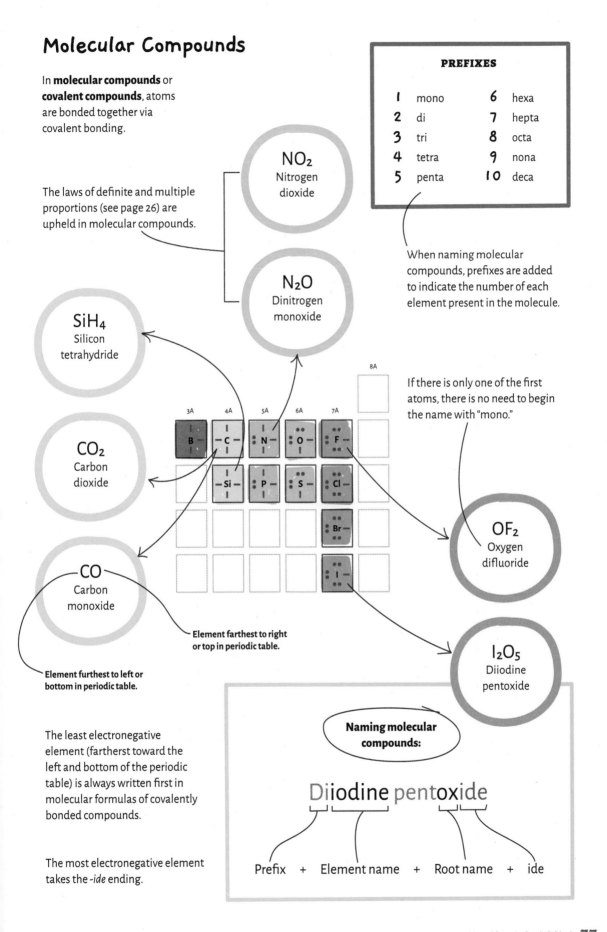

METALLIC BONDING

Metals account for approximately two thirds of all the elements and about 24 percent of the mass of the planet. Whether they are in pure form or mixed with other metals (alloys), metals tend to have high melting temperatures, suggesting that their atoms are bonded together strongly. Metallic bonding is responsible for the bonding between metal atoms, which is quite different to ionic or covalent bonding; the electron sea model is used to explain the nature of metallic bonding.

Metal atoms form a regular pattern when bonded together in three-dimensional space. The resulting atomic arrangement is called a **crystal lattice**.

In a crystal lattice each metal atom is surrounded by other metal atoms.

Metal atoms donate their valence electrons to an **electron sea**, becoming positively charged cations. Think of this as cations swimming in a sea of negatively charged electrons.

All electrons are bound loosely and move constantly through the crystal lattice, holding the cations together strongly.

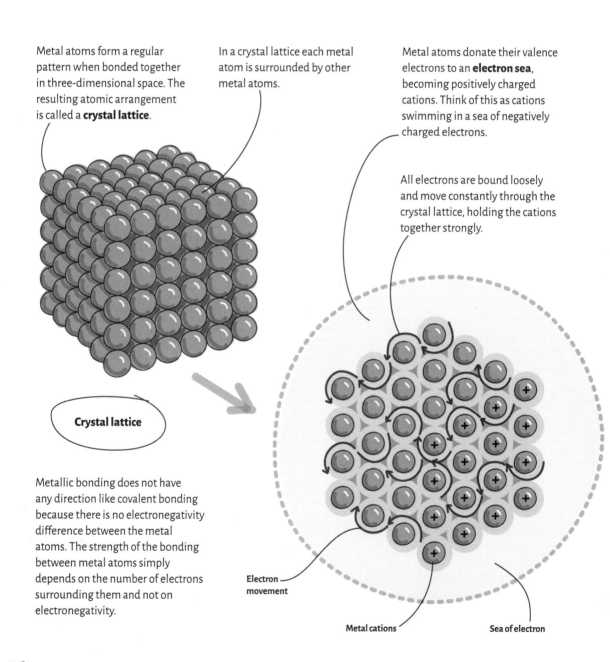

Crystal lattice

Metallic bonding does not have any direction like covalent bonding because there is no electronegativity difference between the metal atoms. The strength of the bonding between metal atoms simply depends on the number of electrons surrounding them and not on electronegativity.

Electron movement

Metal cations

Sea of electron

Luster in Metals

Metals are highly lustrous because they are strong absorbers and reflectors of light. The shiny appearance of metals is due to the high mobility of valence electrons.

Metallic luster of silver

Tarnished silver

Metallic luster of gold

When light (electromagnetic radiation of photons with energy) hits a metallic surface, valence electrons absorb the incoming energy and become excited. As electrons relax back, they emit the absorbed energy as visible light, making metals appear shiny.

Incident visible light

Reflected wave of light

Free metal electrons

Why metals shine

The light that metals reflect is a mixture of all the wavelengths in the visible spectrum, but not in equal proportions. This is why many metals look gray–white in color, but some, such as gold, have a different appearance.

The close packing of metal cations prevents light from passing through, so most of the light is reflected.

How mirrors work

Silver atoms

Light

Transparent glass

Over time, a metallic surface gets dirty, rusty, or oxidized in air, which makes the metal lose its luster; the surface is no longer a pure metal, but a different compound in which electrons are not as free. Silver, for example, gets tarnished as it reacts with oxygen in the air. It needs to be polished to regain its luster.

Mirrors are based on the principle of metals being highly reflective; they are made by coating glass with silver atoms. When light bouncing off an object hits a mirror it is reflected by the silver atoms, making it possible to see a "mirror image."

Black backboard

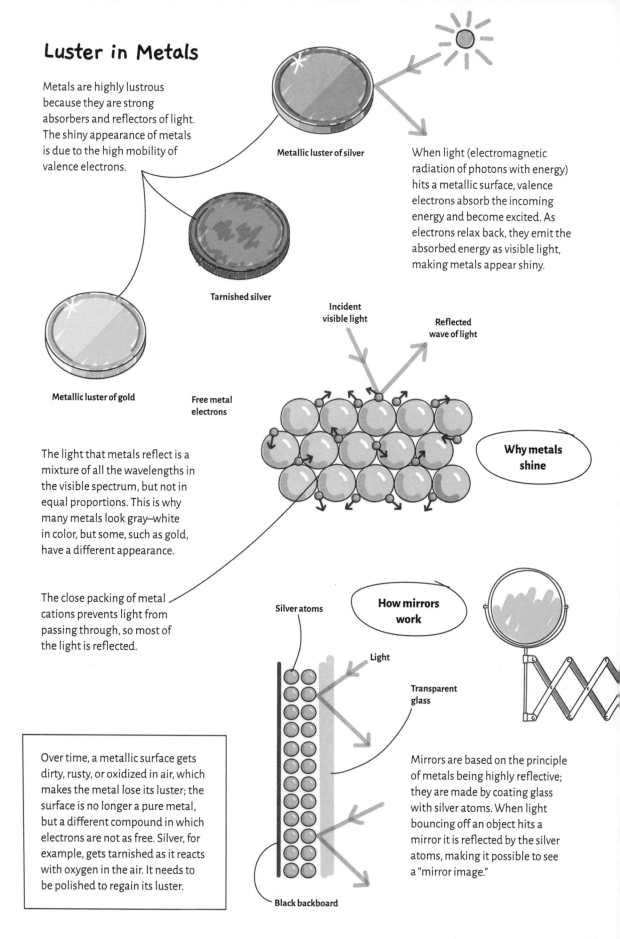

PROPERTIES OF IONIC AND MOLECULAR COMPOUNDS

Covalently bonded molecular compounds and ionic compounds have quite different physical properties. The interaction between ions in an ionic compound is quite strong and uniform throughout the crystal lattice they form. Molecular compounds, on the other hand, display a much wider range of properties. This is because of the variable strength and polarity of their covalent bonds, due to different electronegativities of the bonded atoms. Consequently, molecular compounds can be in gas, liquid, or solid phase at room temperature, while ionic compounds are mostly solid with well-defined crystal shapes.

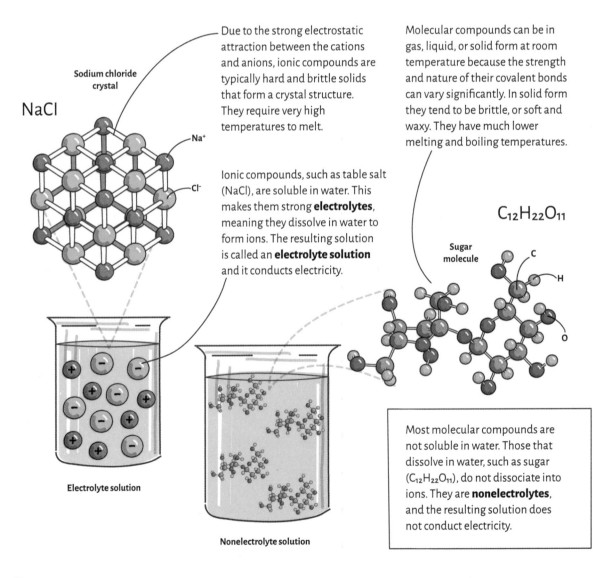

Due to the strong electrostatic attraction between the cations and anions, ionic compounds are typically hard and brittle solids that form a crystal structure. They require very high temperatures to melt.

Molecular compounds can be in gas, liquid, or solid form at room temperature because the strength and nature of their covalent bonds can vary significantly. In solid form they tend to be brittle, or soft and waxy. They have much lower melting and boiling temperatures.

Ionic compounds, such as table salt (NaCl), are soluble in water. This makes them strong **electrolytes**, meaning they dissolve in water to form ions. The resulting solution is called an **electrolyte solution** and it conducts electricity.

Sodium chloride crystal

NaCl

Na$^+$

Cl$^-$

$C_{12}H_{22}O_{11}$

Sugar molecule

C

H

O

Electrolyte solution

Nonelectrolyte solution

Most molecular compounds are not soluble in water. Those that dissolve in water, such as sugar ($C_{12}H_{22}O_{11}$), do not dissociate into ions. They are **nonelectrolytes**, and the resulting solution does not conduct electricity.

Electrolytes and Health

The human body is mostly water, capable of dissolving ionic compounds from essential minerals. Blood circulates electrolytes through the human body for vital biological functions. Electrolytes make it into the body via the food and drink that humans consume.

96.2% of the body weight comes from O, C, H, and N alone. Most of the remaining 3.8% elements are included in electrolytes.

Electrolyte deficiency causes health problems such as heart and muscle issues, anxiety, fatigue, dizziness, insomnia, frequent headaches, and fluid imbalance.

Humans intake electrolytes via a balanced diet consuming various foods and beverages.

Na^+
Sodium

Na^+ maintains fluid balance, nerve function, and muscle contraction.

Cl^-
Chlorine

Cl^- maintains fluid balance.

K^+
Potassium

K^+ regulates heart contraction and maintains fluid balance.

Ca^{2+}
Calcium

Ca^{2+} regulates muscle contraction, nerve function, blood clotting, cell division, and healthy bones and teeth.

Mg^{2+}
Magnesium

Mg^{2+} regulates muscle function, heart rhythm, bone strength, and energy generation.

The human body is 70% water.

 +

COVALENT BONDING

Nonmetal + nonmetal or
nonmetal + metalloid.

IONIC COMPOUND

Forms between a
cation and an anion.

IONIC BONDING

Metal + nonmetal.

TYPES OF CHEMICAL BOND

METALLIC BONDING

Metal + metal.

CHEMICAL BONDING

Hard and brittle solids.

IONIC COMPOUNDS

MOLECULAR COMPOUNDS

Can be gas, liquid, or solid.

PROPERTIES OF IONIC AND MOLECULAR COMPOUNDS

ELECTROLYTE

Forms ions when
dissolved in water.

NONELECTROLYTE

Does not form ions when
dissolved in water.

ESSENTIAL MINERALS

Supply electrolytes to the
human body.

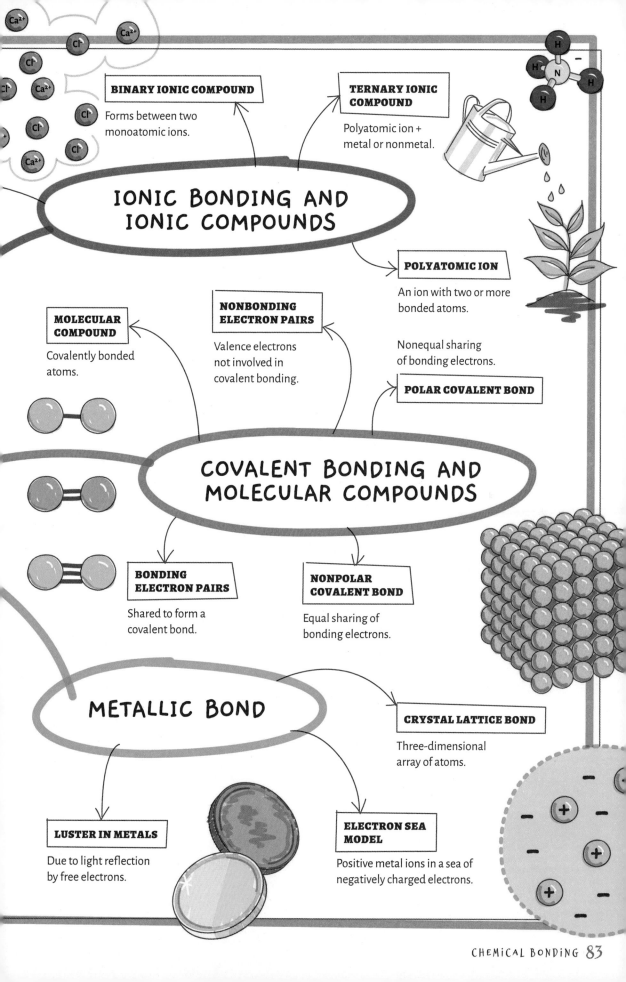

BINARY IONIC COMPOUND

Forms between two monoatomic ions.

TERNARY IONIC COMPOUND

Polyatomic ion + metal or nonmetal.

IONIC BONDING AND IONIC COMPOUNDS

POLYATOMIC ION

An ion with two or more bonded atoms.

MOLECULAR COMPOUND

Covalently bonded atoms.

NONBONDING ELECTRON PAIRS

Valence electrons not involved in covalent bonding.

Nonequal sharing of bonding electrons.

POLAR COVALENT BOND

COVALENT BONDING AND MOLECULAR COMPOUNDS

BONDING ELECTRON PAIRS

Shared to form a covalent bond.

NONPOLAR COVALENT BOND

Equal sharing of bonding electrons.

METALLIC BOND

CRYSTAL LATTICE BOND

Three-dimensional array of atoms.

LUSTER IN METALS

Due to light reflection by free electrons.

ELECTRON SEA MODEL

Positive metal ions in a sea of negatively charged electrons.

CHAPTER 7

MOLECULAR STRUCTURE

There are three main states of matter: solid, liquid, and gas. Which of these states prevails for a substance at room temperature depends on the structure of the particles in it and how these particles interact with each other. The nature and strength of the forces of interaction between the molecules in a substance—**intermolecular forces**—play a central role in the observed properties of matter. In any molecular compound, the intermolecular forces are determined primarily by the geometric shape and polarity of the molecules; water, for example, would not exist without the polar nature of the water molecule, with its bent geometric shape.

LEWIS DOT STRUCTURES OF MOLECULAR COMPOUNDS

Lewis dot structures are two-dimensional representations of molecules that make it easier to visualize the distribution of valence electrons, whether they exist as lone pairs or as part of a covalent bond. The visual information that Lewis dot structures provide can be used to predict the polarity of a molecule, as well as the type and strength of its intermolecular forces. Ice floats because it is less dense than liquid water—a phenomenon that can be deduced directly from the Lewis structure of a water molecule.

Lewis dot structures are built from skeletal figures of molecules, with the least electronegative element (the **central atom**) in the middle. Covalent bonds connect the terminal atoms to the central atom.

There may be more than one central atom, in which case terminal atoms are distributed evenly around the central atoms when forming the skeletal structure.

A Lewis dot structure shows a two-dimensional image of a molecule, demonstrating the valence electron distribution for the molecule.

STEP 1:
Write the skeletal structure with the least electronegative element in the middle as the central atom.

STEP 2:
Write the Lewis dot symbols for each atom and start forming covalent bonds to the central atom.

STEP 3:
Show covalent bonds and nonbonding electron pairs.

STEP 4:
Form double or triple bonds when necessary for an octet.

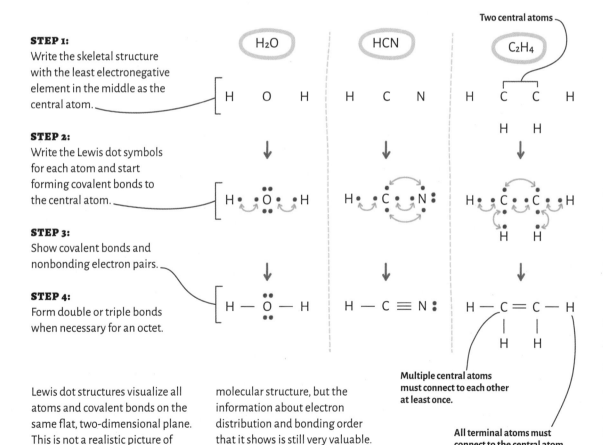

Multiple central atoms must connect to each other at least once.

All terminal atoms must connect to the central atom at least once.

Lewis dot structures visualize all atoms and covalent bonds on the same flat, two-dimensional plane. This is not a realistic picture of molecular structure, but the information about electron distribution and bonding order that it shows is still very valuable.

VSEPR THEORY: MOLECULAR GEOMETRY

The electrostatic repulsion between negatively charged electrons forms the basis of how covalent bonds are arranged in a molecule. **Valence shell electron pair repulsion (VSEPR)** theory is based on the idea that **electron groups**—defined as lone pairs, single bonds, double bonds, or triple bonds—repel one another. The theory focuses on the repulsive forces between the electron groups around each central atom in a molecule, and on the specific three-dimensional geometries that emerge as a result of the electron group arrangements.

Electron Group Geometry

The number of electron groups around the central atom determines the maximum separation between them due to electrostatic repulsions.

There are three types of repulsion that push these electron groups away from each other: **bonding pair–bonding pair**, **lone pair–bonding pair**, and **lone pair–lone pair**.

Definite geometric shapes around the central atom emerge, demonstrating the angles between the electron groups around the central atom.

180°

Lone pair–bonding pair repulsion.

109.5°

Linear electron group geometry emerges when there are only two electron groups on the central atom, separated by 180 degrees.

Bonding pair–bonding pair repulsion.

Lone pair–lone pair repulsion.

Tetrahedral electron group geometry emerges when there are four electron groups on the central atom, separated by 109.5 degrees.

Trigonal planar electron group geometry emerges when there are three electron groups on the central atom, separated by 120 degrees.

120°

Molecular Shape

The linear, trigonal planar, and tetrahedral geometries emerge as the shape of a molecule if all of the electron groups around the central atom are bonding pairs. When lone pairs are present on the central atom, the molecular shape gets modified due to **bond angle distortion**. Lone electron pairs, with more freedom to move around, push down on bonding pairs, causing a reduction in the ideal bond angles. No bond angle distortion is observed for molecules with linear geometry.

Electron group geometry is tetrahedral with no bond angle distortion because there are no lone pairs on the central atom.

Molecular shape is also tetrahedral with ideal bond angles because all electron groups are bonding to other atoms.

Electron group geometry is tetrahedral but bond angles are reduced due to the lone pair on the central atom.

Only three electron groups are bonding and the molecular shape is trigonal pyramidal.

Electron group geometry is tetrahedral but bond angles are reduced due to two lone pairs on the central atom.

Only two electron groups are bonding with a bent molecular shape.

Electron group geometry is trigonal planar but bond angles are reduced due to the lone pair on the central atom.

Only two electron groups are bonding with a bent molecular shape.

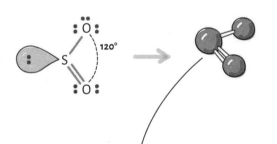

The actual shape of a molecule is determined solely by the geometry of the bonding electron pairs (covalent bonds).

As the number of lone pairs around the central atom rises, the bond angle distortion increases as well.

MOLECULAR SHAPE AND POLARITY

Molecular compounds can have **polar** or **nonpolar** nature depending upon their valence electron distribution. Molecular geometry reveals how valence electrons are scattered around all the atoms in a molecule. If electrons are distributed about the molecule unevenly, molecular regions with more and less electron density emerge, which makes the molecule polar. Polar molecules are also referred to as **dipoles**. A uniform valence electron distribution, on the other hand, creates a nonpolar molecule.

Molecules with lone electron pairs on the central atom tend to be polar because they cause bond angle distortion and a nonuniform electron distribution.

Molecules with no lone pairs on the central atom tend to be nonpolar if all the terminal atoms are identical. If the terminal atoms are not the same, possessing different electronegativities, a polar molecule is formed.

Highly electronegative oxygen draws an electron toward itself.

A stream of polar water molecules gets bent by a charged rod due to the presence of a molecular dipole.

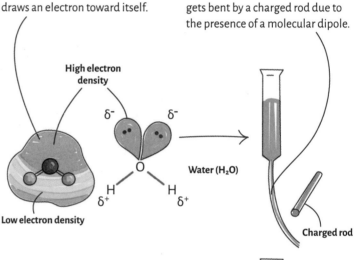

High electron density

Low electron density

Water (H_2O)

Charged rod

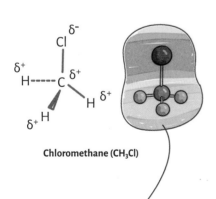

Chloromethane (CH_3Cl)

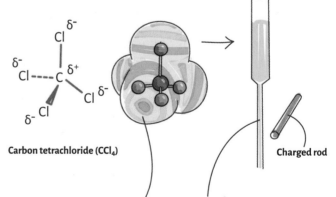

Carbon tetrachloride (CCl_4)

Charged rod

Highly electronegative chlorine causes an uneven electron distribution around the molecule creating a dipole.

Identical terminal atoms around the central atom results in uniform electron distribution.

Nonpolar molecules with no dipole do not respond to external charges.

Like Dissolves Like

Polar and nonpolar molecules have different properties. Water is a highly polar substance, making it capable of attracting or mixing with other polar substances. This is known as the **like dissolves like principle**. Water alone cannot be used to kill the coronavirus because the surface of the virus is made up of nonpolar fat molecules.

Soap molecules, called **surfactants**, consist of a polar head and a nonpolar tail. In water, soap molecules bunch together to form large molecules known as **micelles**, which break up during washing, releasing surfactant molecules capable of bonding polar water molecules to nonpolar fat molecules.

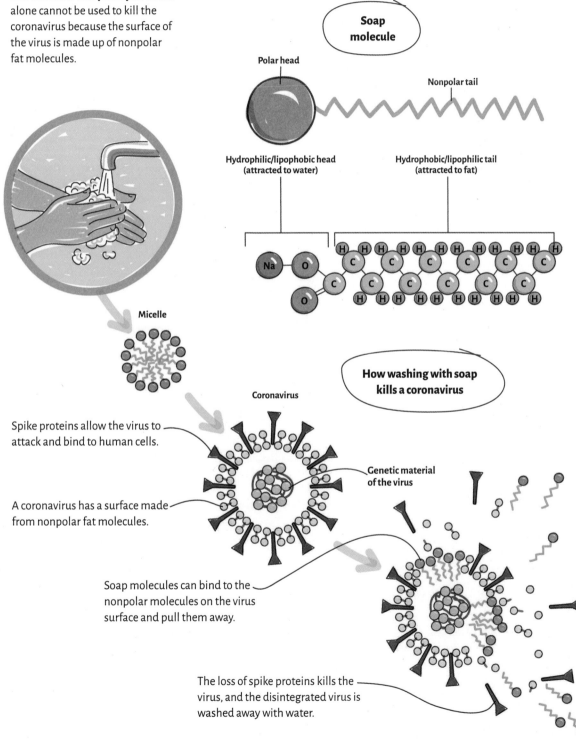

Soap molecule

Polar head

Nonpolar tail

Hydrophilic/lipophobic head
(attracted to water)

Hydrophobic/lipophilic tail
(attracted to fat)

Micelle

How washing with soap kills a coronavirus

Coronavirus

Spike proteins allow the virus to attack and bind to human cells.

Genetic material of the virus

A coronavirus has a surface made from nonpolar fat molecules.

Soap molecules can bind to the nonpolar molecules on the virus surface and pull them away.

The loss of spike proteins kills the virus, and the disintegrated virus is washed away with water.

iNTERMOLECULAR FORCES

ntramolecular forces—ionic, covalent, and metallic bonds—involve valence electrons. When molecules of a compound held together by intramolecular forces get sufficiently close to one another, attractive forces known as **van der Waals forces** start to act among them. These are **intermolecular forces**, which do not involve the sharing of valence electrons. Intermolecular forces are much weaker than intramolecular forces, but they play an important role in the physical properties of matter.

Strength of Intermolecular Forces

In water, the intramolecular forces (covalent bonds) are what give the molecule its bent shape, but intermolecular forces also exist, due to the attraction between the oppositely charged regions of water's polar molecules.

The polarity of a molecule determines the type and strength of its intermolecular forces. For solids, molecules and atoms are in close contact as a result of strong intermolecular forces.

Gas molecules are largely separated from each other, indicating weak intermolecular forces.

Liquid molecules are in contact but have mobility because of the intermediate strength of their intermolecular forces.

The addition of heat weakens intermolecular forces, so phase changes from solid to liquid (melting) and from liquid to gas (vaporization) can take place.

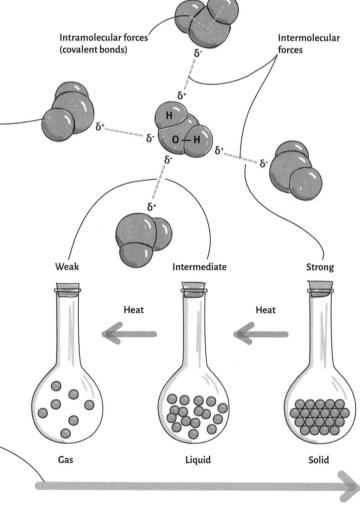

Intramolecular forces (covalent bonds)

Intermolecular forces

δ^-

δ^+

H

O—H δ^+ .... δ^-

δ^+ .... δ^-

δ^-

δ^+

Weak   Intermediate   Strong

Heat   Heat

Gas   Liquid   Solid

Weak intermolecular forces   Strong intermolecular forces

Types of Intermolecular Forces

London dispersion forces are quite weak. They exist in all types of molecules, but are the only forces between the molecules of a nonpolar substance.

The valence electron distribution in nonpolar molecules is pretty even, so they have no dipole.

A temporary, short-lived dipole can occur in nonpolar molecules when they get close to other molecules. This **induced dipole** can interact with other induced dipoles in the sample, creating weak, short-lived London dispersion forces.

These forces break very easily. In fact, many nonpolar substances are gases at room temperature because London dispersion forces are not strong enough to keep the molecules close enough for a liquid or solid state.

Dipole-dipole forces exist between the molecules of a polar substance with permanent dipoles. This is due to the nonuniform distribution of valence electrons.

Attractions and repulsions occur between the positively and negatively charged regions of molecules, but attractions typically dominate. Dipole-dipole forces are much stronger than London dispersion forces.

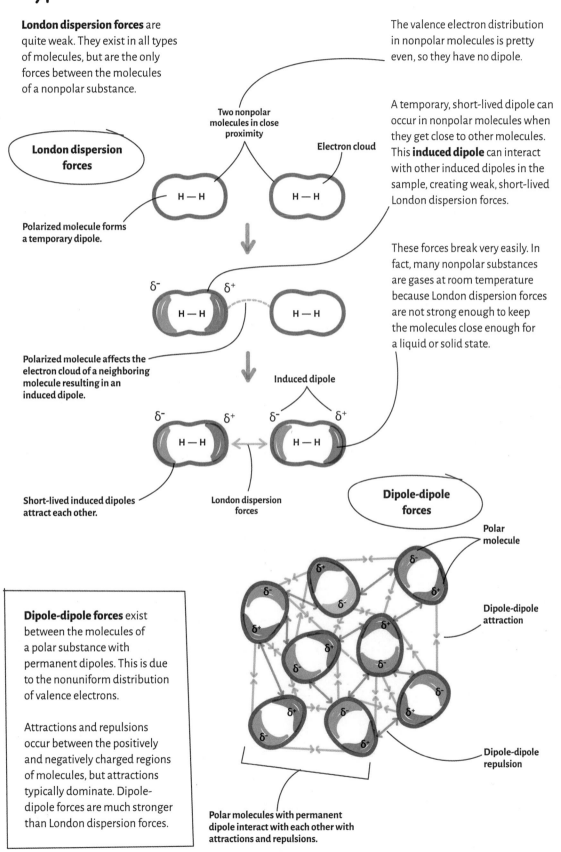

London dispersion forces

Two nonpolar molecules in close proximity

Electron cloud

H — H   H — H

Polarized molecule forms a temporary dipole.

δ^- δ^+

H — H   H — H

Polarized molecule affects the electron cloud of a neighboring molecule resulting in an induced dipole.

Induced dipole

δ^- δ^+ δ^- δ^+

H — H   H — H

Short-lived induced dipoles attract each other.

London dispersion forces

Dipole-dipole forces

Polar molecule

Dipole-dipole attraction

Dipole-dipole repulsion

Polar molecules with permanent dipole interact with each other with attractions and repulsions.

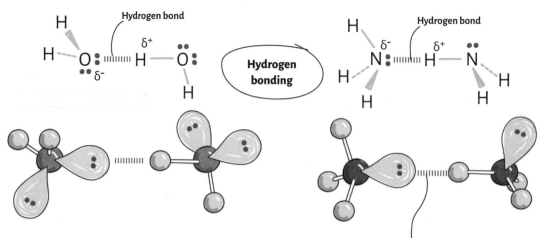

Hydrogen bonding

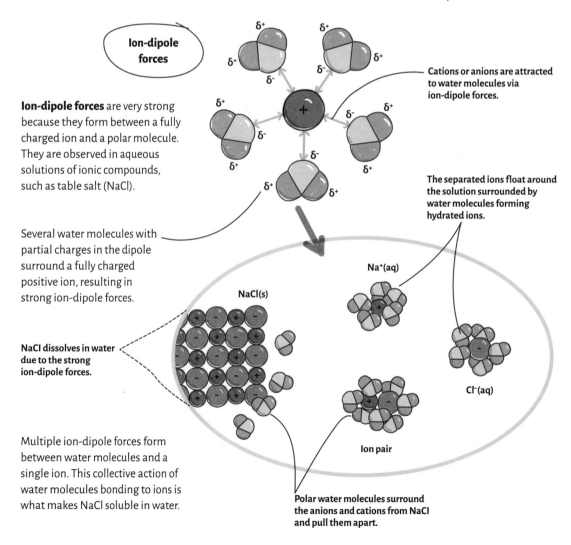

Hydrogen bonding is a dipole-dipole force, but because of its unusual strength it is studied separately. Hydrogen bonding occurs between polar molecules that have an H atom directly connected to an F, O, or N atom. Hydrogen bonding does not form if these H–F, H–O, or H–N covalent bonds are not present in the molecules.

Hydrogen bonding forms between the dipoles of polar molecules. The bonds are stronger than other dipole-dipole forces, so many substances with hydrogen bonding (such as water) tend to be liquids at room temperature.

Ion-dipole forces

Ion-dipole forces are very strong because they form between a fully charged ion and a polar molecule. They are observed in aqueous solutions of ionic compounds, such as table salt (NaCl).

Cations or anions are attracted to water molecules via ion-dipole forces.

Several water molecules with partial charges in the dipole surround a fully charged positive ion, resulting in strong ion-dipole forces.

The separated ions float around the solution surrounded by water molecules forming hydrated ions.

NaCl dissolves in water due to the strong ion-dipole forces.

Multiple ion-dipole forces form between water molecules and a single ion. This collective action of water molecules bonding to ions is what makes NaCl soluble in water.

NaCl(s)

Na$^+$(aq)

Cl$^-$(aq)

Ion pair

Polar water molecules surround the anions and cations from NaCl and pull them apart.

Hydrogen Bonding in Action

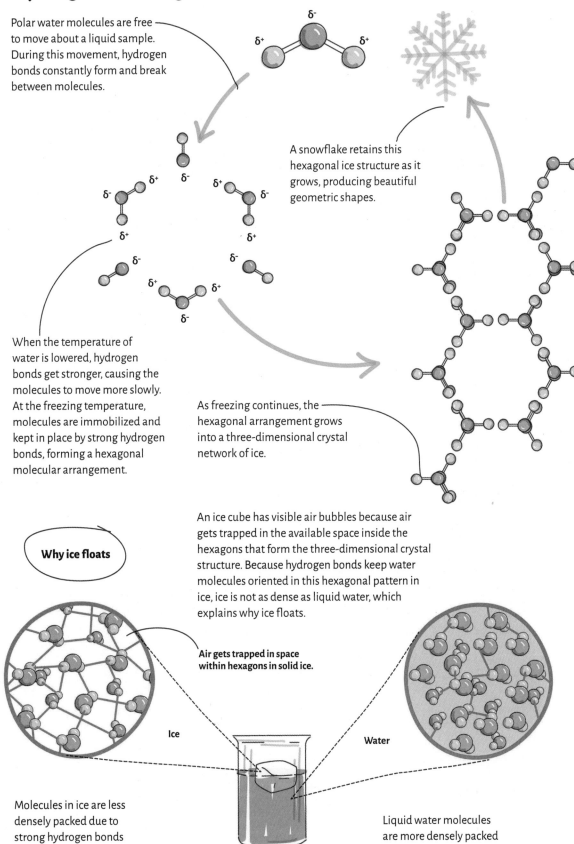

Polar water molecules are free to move about a liquid sample. During this movement, hydrogen bonds constantly form and break between molecules.

A snowflake retains this hexagonal ice structure as it grows, producing beautiful geometric shapes.

When the temperature of water is lowered, hydrogen bonds get stronger, causing the molecules to move more slowly. At the freezing temperature, molecules are immobilized and kept in place by strong hydrogen bonds, forming a hexagonal molecular arrangement.

As freezing continues, the hexagonal arrangement grows into a three-dimensional crystal network of ice.

An ice cube has visible air bubbles because air gets trapped in the available space inside the hexagons that form the three-dimensional crystal structure. Because hydrogen bonds keep water molecules oriented in this hexagonal pattern in ice, ice is not as dense as liquid water, which explains why ice floats.

Why ice floats

Air gets trapped in space within hexagons in solid ice.

Ice

Water

Molecules in ice are less densely packed due to strong hydrogen bonds

Liquid water molecules are more densely packed

BONDING FORCES AND CRYSTALLINE SOLIDS

Crystalline solids are described by the types of particles in them and the types of forces that keep those particles together. **Interparticle forces** (intermolecular, interionic, and interatomic forces) have a different strength and bonding structure. Crystalline solids exhibit physical properties compatible with the type and strength of the forces that keep their particles together.

Hydrogen bond

Hydrogen bond

Ice:
molecular solid

Based on the individual particles present, crystalline solids are classified into three main categories: **molecular solids**, **ionic solids**, and **atomic solids**. Atomic solids are categorized into **nonbonded solids**, **metallic solids**, and **network covalent solids**, depending on the type of forces that exist between the atoms.

Molecules make up molecular solids, which are held together by intermolecular forces. They tend to be soft and melt at low temperatures.

Network covalent bond

Diamond:
network covalent
atomic solid

Atomic solids in which the atoms are bonded together by strong covalent bonds are known for their extreme hardness, durability, and high melting temperatures. They make up some of the strongest known natural substances, such as diamond. The three-dimensional array of carbon atoms in a diamond are held together by very strong and directional covalent bonds.

Carbon atoms

Unit cell

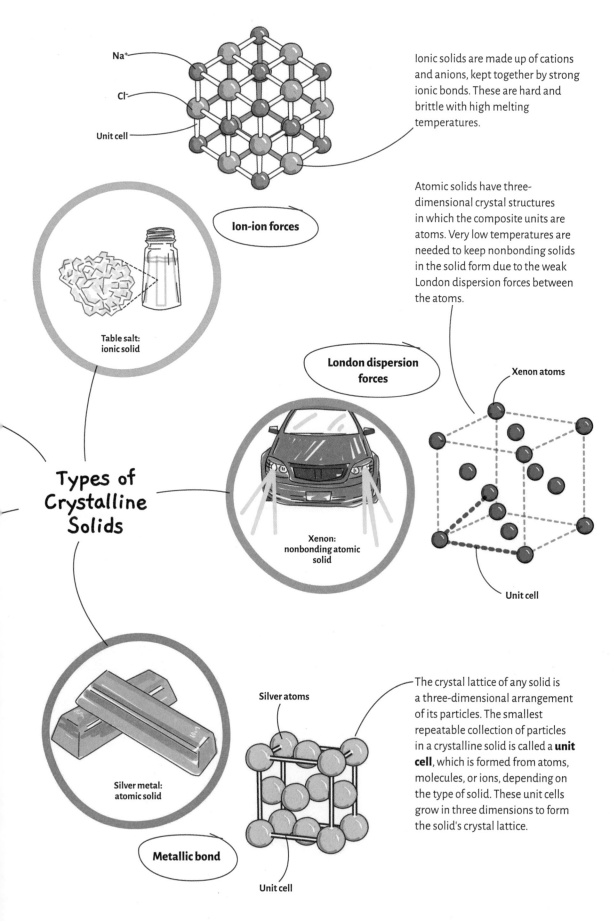

Na⁺

Cl⁻

Unit cell

Ionic solids are made up of cations and anions, kept together by strong ionic bonds. These are hard and brittle with high melting temperatures.

Ion-ion forces

Table salt: ionic solid

Atomic solids have three-dimensional crystal structures in which the composite units are atoms. Very low temperatures are needed to keep nonbonding solids in the solid form due to the weak London dispersion forces between the atoms.

London dispersion forces

Xenon atoms

Types of Crystalline Solids

Xenon: nonbonding atomic solid

Unit cell

Silver atoms

Silver metal: atomic solid

The crystal lattice of any solid is a three-dimensional arrangement of its particles. The smallest repeatable collection of particles in a crystalline solid is called a **unit cell**, which is formed from atoms, molecules, or ions, depending on the type of solid. These unit cells grow in three dimensions to form the solid's crystal lattice.

Metallic bond

Unit cell

H :Ö: H

The least electronegative atom in a molecule.

CENTRAL ATOM

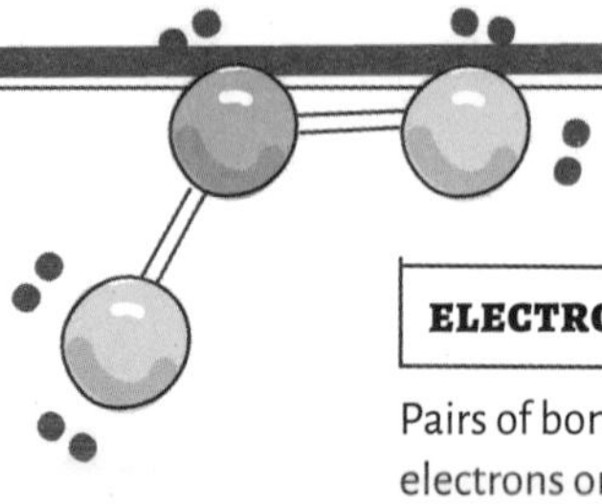

ELECTRON GROUP

Pairs of bonding and nonbonding electrons on a central atom.

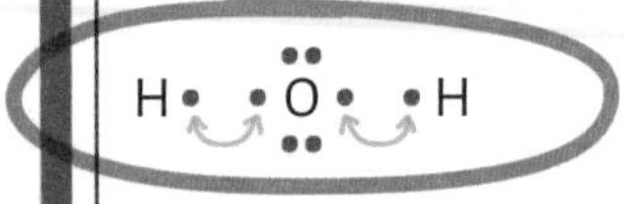

LINEAR SHAPE

Two covalent bonds.

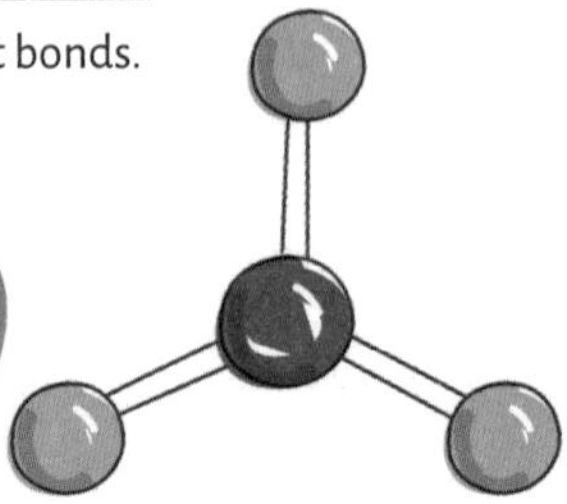

LEWIS DOT STRUCTURES OF MOLECULAR COMPOUNDS

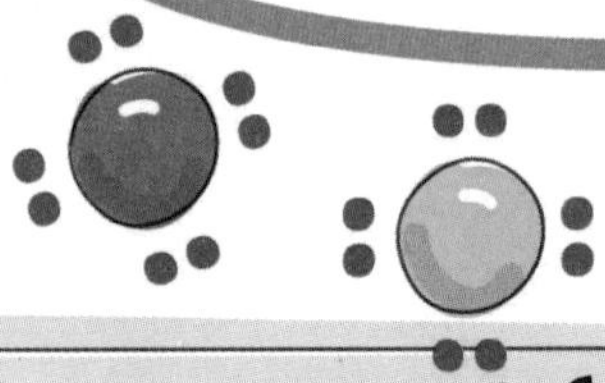

TRIGONAL PLANAR SHAPE

Three covalent bonds.

MOLECULAR STRUCTURE

INTERPARTICLE FORCES

Forces that keep atoms, molecules, and ions together in solids.

The smallest repeatable collection of particles in crystals.

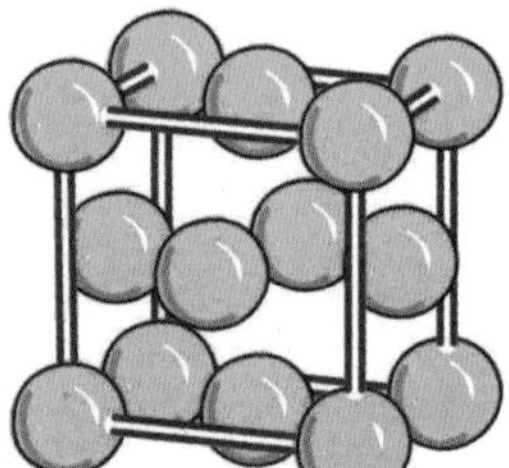

UNIT CELL

BONDING FORCES AND CRYSTALLINE SOLIDS

IONIC SOLIDS

Made of cations and anions.

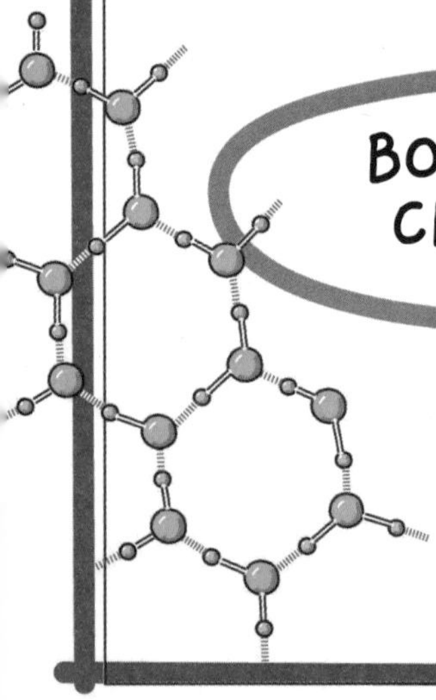

MOLECULAR SOLIDS

Made of molecules.

ATOMIC SOLIDS

Made of atoms.

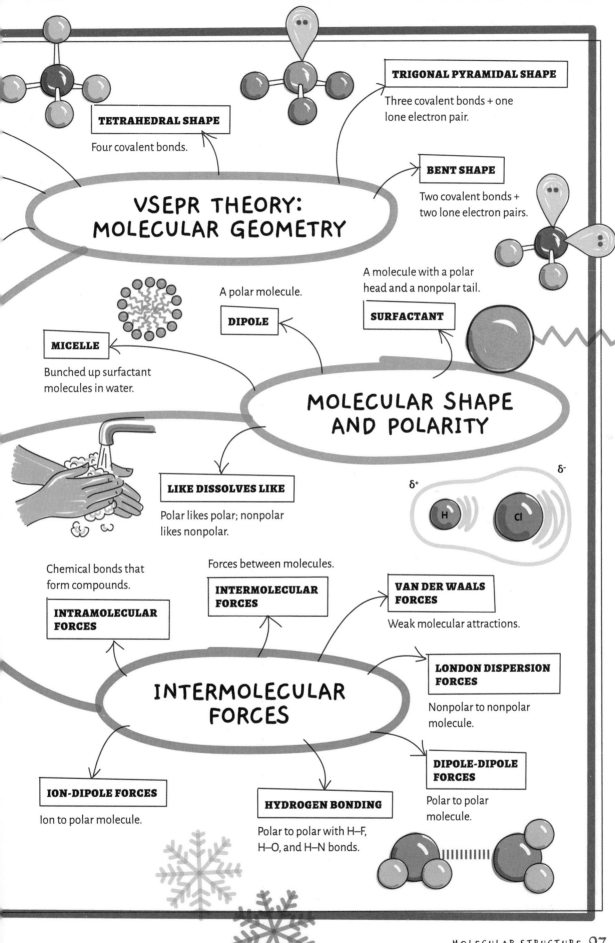

TETRAHEDRAL SHAPE

Four covalent bonds.

TRIGONAL PYRAMIDAL SHAPE

Three covalent bonds + one lone electron pair.

BENT SHAPE

Two covalent bonds + two lone electron pairs.

VSEPR THEORY: MOLECULAR GEOMETRY

A molecule with a polar head and a nonpolar tail.

SURFACTANT

A polar molecule.

DIPOLE

MICELLE

Bunched up surfactant molecules in water.

MOLECULAR SHAPE AND POLARITY

δ^+ δ^-

H   Cl

LIKE DISSOLVES LIKE

Polar likes polar; nonpolar likes nonpolar.

Chemical bonds that form compounds.

INTRAMOLECULAR FORCES

Forces between molecules.

INTERMOLECULAR FORCES

VAN DER WAALS FORCES

Weak molecular attractions.

LONDON DISPERSION FORCES

Nonpolar to nonpolar molecule.

INTERMOLECULAR FORCES

DIPOLE-DIPOLE FORCES

Polar to polar molecule.

ION-DIPOLE FORCES

Ion to polar molecule.

HYDROGEN BONDING

Polar to polar with H–F, H–O, and H–N bonds.

CHAPTER 8

CHEMICAL REACTIONS AND STOICHIOMETRY

A **chemical reaction** is a process in which chemical changes of matter take place. A chemical reaction does not change the identity of the atoms involved; it simply rearranges the constituent atoms in one or more substances (reactants) to form one or more different substances (products). For this to occur, chemical bonds in the reactant molecules break, and new chemical bonds form simultaneously in the product molecules. In this way, a chemical reaction is essentially a recipe that provides the instructions for making a new substance. Chemists use the mole concept (see page 31) and carry out stoichiometry calculations to determine the amounts of reactants and products for a correctly written and balanced chemical reaction.

WRITING AND BALANCING CHEMICAL EQUATIONS

Chemical reactions provide instructions about which reactants to use to create the products of interest. A universal language with chemical symbols of atoms, molecules, and ions is used to represent a chemical change in a **chemical equation**, and all of these chemical symbols must be provided correctly. Additionally, a chemical equation must be properly balanced to uphold the law of conservation of mass.

Chemical Equations

In chemical equations, an arrow—meaning **yield** or **produce**—separates the reactants and products. Multiple reactants and/or products are separated by a plus sign.

It helps scientists to know the physical state of each reactant and product. Therefore, lowercase letters are used to symbolize the solid (s), liquid (l), gas (g), and dissolved in water/aqueous (aq) forms; this is written next to each substance.

Color change

Temperature change

+ sign separates multiple reactants and products

Yield

Physical state of substance:

g: gas
s: solid
l: liquid
aq: aqueous

$$C\ (s) + O_2\ (g) \longrightarrow CO_2\ (g)$$

Reactants

Product

Oxygen gas, O_2, from air

Indicators of chemical change

Carbon dioxide gas, CO_2

Solid carbon, C

Chemical change

Precipitate formation

Gas formation

Depending on the type of reaction, there are different forms of evidence that indicate that a chemical change has taken place. The most common evidence of chemical change include color and temperature changes, or the formation of light, sound, gas bubbles, or a solid precipitate. Several of these "clues" may be present in the same reaction.

Light formation

Balancing Chemical Equations

Baking a delicious cake requires carefully measured amounts of all ingredients that must be combined in the proper ratios as provided by the recipe. The same idea is valid for chemical reactions and chemists use stoichiometry to make sure all reaction components are prepared in correct proportions.

3 eggs

+

250 ml milk

=

+

5 pancakes (1 batch)

125 g flour

Just like a recipe, chemical equations provide information about which reactants (ingredients) need to be used in order to produce the desired products. The measuring unit in a chemical equation is the mole, which appears as a whole number in front of each reactant and product. These numbers are called **stoichiometric coefficients**.

A recipe provides the correct amounts of each ingredient (reactant).

Desired product can only be made if the recipe is followed properly.

Accurate proportions of reactants are needed to make the desired product.

$H_2 (g)$ + $I_2 (g)$ → $2HI (g)$

1 mol H_2 1 mol I_2 2 moles HI

2.016 g H_2 253.8 g I_2 255.816 g HI

Stoichiometric coefficient provides mole numbers for balancing.

A chemical equation must uphold the law of conservation of mass, so the total mass of the reactants is equal to the total mass of the products. Correctly placed

stoichiometric coefficients balance a chemical equation so that each type of atom appears in equal numbers on both sides of the equation.

Stoichiometric coefficients provide amounts in moles, which can be converted into other units for convenient measurement.

Rules for Balancing Chemical Equations

A chemical reaction is translated from words to a skeleton chemical equation showing the reactants and desired products. This equation simply shows what compounds are reacting, and what products are produced.

The law of conservation of mass is satisfied by using appropriate stoichiometric coefficients in the skeleton equation to achieve a balanced reaction equation that shows the correct reactants and desired products.

In balancing, subscripted numbers cannot be changed, as that would change the identity of a reactant or product.

New reactants or products cannot be added in order to balance a chemical equation.

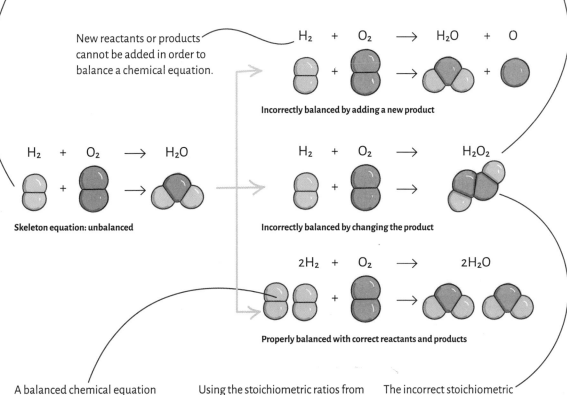

$H_2 + O_2 \longrightarrow H_2O + O$

Incorrectly balanced by adding a new product

$H_2 + O_2 \longrightarrow H_2O$

Skeleton equation: unbalanced

$H_2 + O_2 \longrightarrow H_2O_2$

Incorrectly balanced by changing the product

$2H_2 + O_2 \longrightarrow 2H_2O$

Properly balanced with correct reactants and products

A balanced chemical equation provides information about the necessary ratios between reactants and products.

Using the stoichiometric ratios from the balanced chemical equation, chemists can produce any amount of the desired product.

The incorrect stoichiometric ratio between hydrogen gas and oxygen gas could produce hydrogen peroxide (H_2O_2) instead of water (H_2O). These are two very different products!

2:1 ratio is needed to make stoichiometric amount of product.

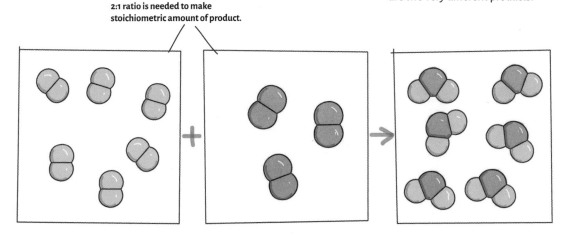

STOICHIOMETRIC RELATIONSHIPS

The coefficients in a balanced chemical reaction establish the mole ratios between the substances involved in the reaction. These stoichiometric ratios can be used to determine how much of the reactant substances are needed to make the desired amount of product. The study of the numerical relationships between the amounts of reactants used and products formed is called **reaction stoichiometry**. Chemists routinely carry out such calculations to plan and execute chemical reactions that yield the desired amount of product.

Stoichiometry

Coefficients in a balanced chemical equation show the stoichiometric ratios in moles between each substance on the reactant and product sides. These mole ratios are used to determine how much reactant would be needed to produce the desired amount of product.

Sodium azide (NaN_3) in powder form is the reactant in the chemical process that produces the nitrogen gas needed to fill the airbag in an automobile. Engineers determine the precise volume of nitrogen gas needed for a certain airbag, and from that, chemists can carry out stoichiometric calculations to determine what mass of sodium azide is required.

The balanced equation for the reaction in an airbag calls for 2 mol of sodium azide, but this is not necessarily the amount needed, because it may not generate the required volume of nitrogen gas. Stoichiometric ratios enable chemists to determine the necessary moles of any substance in a chemical reaction. Conversion factors make it possible to convert moles to any other convenient unit of measurement, such as mass and volume.

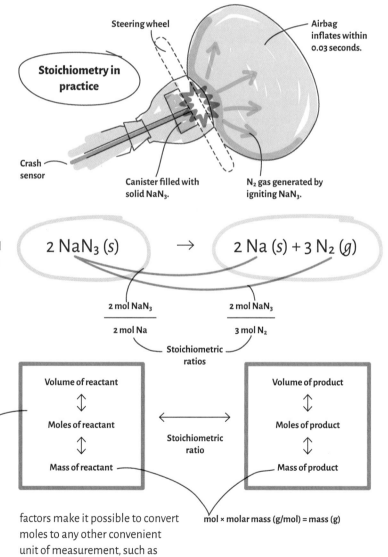

Stoichiometry in practice

Steering wheel

Airbag inflates within 0.03 seconds.

Crash sensor

Canister filled with solid NaN_3.

N_2 gas generated by igniting NaN_3.

$$2 \ NaN_3 \ (s) \quad \rightarrow \quad 2 \ Na \ (s) + 3 \ N_2 \ (g)$$

$\dfrac{2 \ mol \ NaN_3}{2 \ mol \ Na}$ $\dfrac{2 \ mol \ NaN_3}{3 \ mol \ N_2}$

Stoichiometric ratios

Volume of reactant
↕
Moles of reactant
↕
Mass of reactant

Stoichiometric ratio

Volume of product
↕
Moles of product
↕
Mass of product

mol × molar mass (g/mol) = mass (g)

Limiting Reagent

Chemists do not always have stoichiometric amounts of reactants available, which limits the amount of product they can make from a chemical reaction. The reactant that runs out first in a reaction is the **limiting reagent**. This determines how much product the reaction will yield.

From three bodies and eight tires, only two complete trucks can be made before the supply of tires (the limiting reagent) runs out, preventing the manufacture of more trucks. There will be one body left from this process, which is referred to as **excess reagent**.

The concept of limiting reagent is important in chemistry. It tells us the maximum amount of product (the **theoretical yield**) we can make from a reaction before the limiting reagent is depleted. Knowing the theoretical yield enables us to adjust the amounts of reactants to prevent chemical waste.

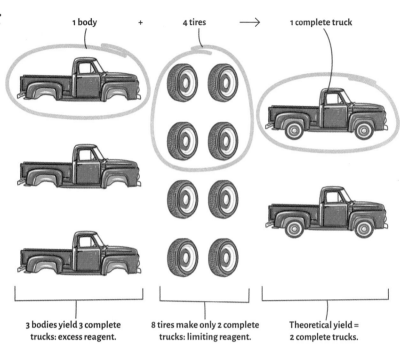

1 body + 4 tires → 1 complete truck

3 bodies yield 3 complete trucks: excess reagent.

8 tires make only 2 complete trucks: limiting reagent.

Theoretical yield = 2 complete trucks.

When 10 hydrogen molecules react with 10 oxygen molecules, only 10 water molecules can be produced before the hydrogen runs out, due to the 2:1 stoichiometric ratio between the reactant molecules. Therefore, the theoretical yield for this particular reaction is 10 water molecules.

Some of the reaction products can be lost due to experimental error, reducing the amount of product we actually make (the actual yield). For example, if we have an actual yield of 8 water molecules, from a theoretical yield of 10 water molecules, we end up with 2 fewer water molecules than the limiting reagent allows. In this instance the reaction yield is 80 percent.

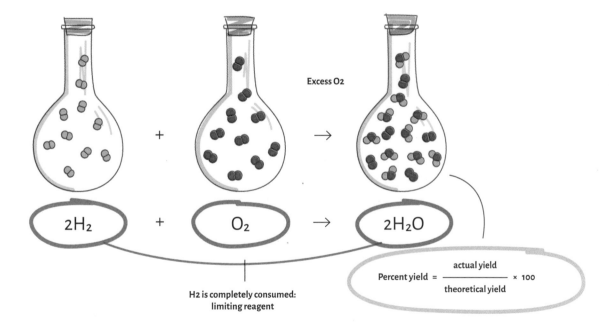

Excess O_2

$$2H_2 + O_2 \rightarrow 2H_2O$$

H_2 is completely consumed: limiting reagent

$$\text{Percent yield} = \frac{\text{actual yield}}{\text{theoretical yield}} \times 100$$

TYPES OF CHEMICAL REACTIONS

All chemical reactions involve a transformation of reactants into products without changing the identity of the atoms involved. However, once initiated, they may proceed in different ways. A general classification of chemical reactions is achieved by considering some of the mechanisms that take place as atoms in reactants are rearranged to form the product.

In **redox** (reduction and oxidation) **reactions**, a transfer of electrons takes place between two different atoms in the reactants. An **oxidation** process takes place when an atom loses electrons while **reduction** involves an atom gaining electrons. Redox reactions are utilized in batteries to generate electrical power.

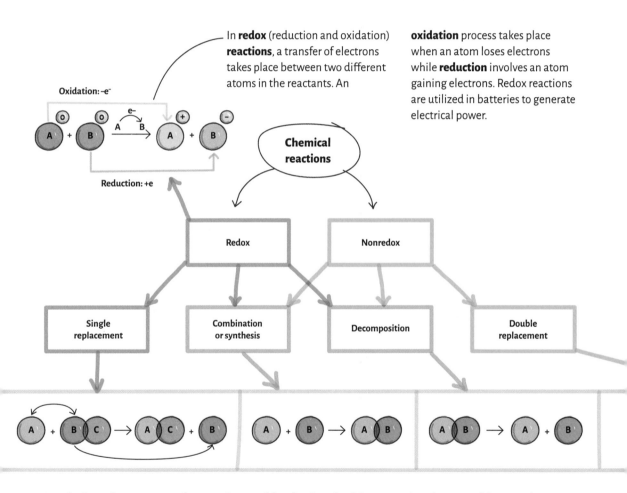

In a **single replacement reaction,** one element replaces a similar element in a compound.

In a **combination (synthesis) reaction**, two or more compounds combine to form a new substance; hydrogen gas and oxygen gas undergo a combination reaction to form water, for example.

In a **decomposition reaction,** a single compound decomposes or breaks down into two or more simpler substances. For example, the decomposition reaction of sodium azide (NaN_3) into elemental sodium and nitrogen gas is utilized in automobile airbags.

We encounter many different types of chemical reactions in our daily lives. Some of these reactions are essential for life, while others are needed for convenience and enjoyment.

Photosynthesis reactions provide plants with the food they need to survive, while also generating oxygen for fresher air. Sunlight catalyzes the reaction between atmospheric carbon dioxide and water to form glucose ($C_6H_{12}O_6$).

Combustion reactions of fuels provide heat and mechanical energy. A hydrocarbon compound, consisting of carbon and hydrogen, reacts with oxygen gas to produce water, carbon dioxide, and heat. Octane (C_8H_{18}) is the main component of gasoline and its combustion provides the necessary power to run a vehicle.

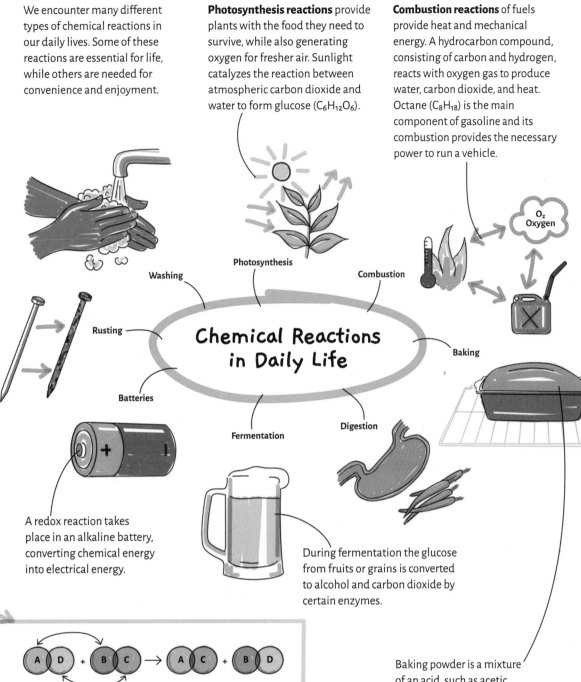

A redox reaction takes place in an alkaline battery, converting chemical energy into electrical energy.

During fermentation the glucose from fruits or grains is converted to alcohol and carbon dioxide by certain enzymes.

In a **double replacement reaction**, the cations and anions in two ionic compounds trade places to form two new ionic compounds. These reactions typically occur in aqueous solutions.

Baking powder is a mixture of an acid, such as acetic acid ($C_2H_4O_2$), and sodium bicarbonate ($NaHCO_3$). When the wet and dry ingredients are mixed together during baking, a reaction takes place that produces carbon dioxide, which causes baked goods to rise.

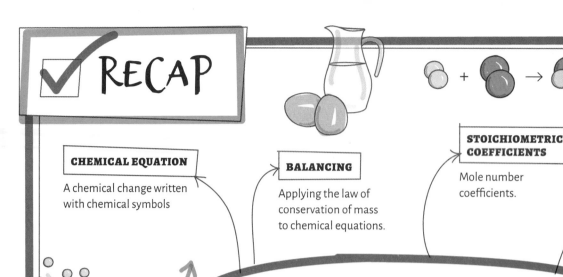

CHEMICAL EQUATION

A chemical change written with chemical symbols

BALANCING

Applying the law of conservation of mass to chemical equations.

STOICHIOMETRIC COEFFICIENTS

Mole number coefficients.

O_2

CO_2

C

WRITING AND BALANCING CHEMICAL EQUATIONS

CHEMICAL REACTIONS AND STOICHIOMETRY

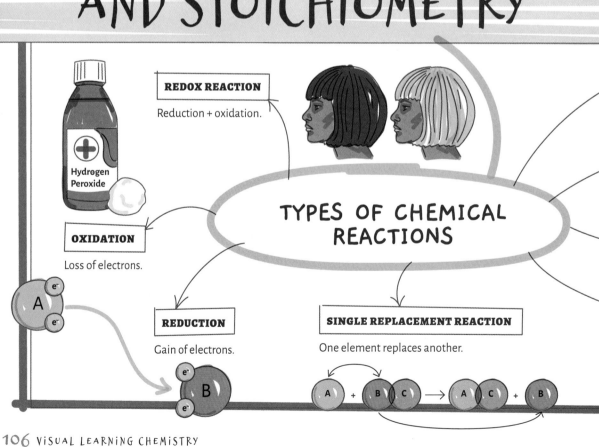

REDOX REACTION

Reduction + oxidation.

Hydrogen Peroxide

OXIDATION

Loss of electrons.

TYPES OF CHEMICAL REACTIONS

A e^- e^-

REDUCTION

Gain of electrons.

e^- B e^-

SINGLE REPLACEMENT REACTION

One element replaces another.

A + B C → A C + B

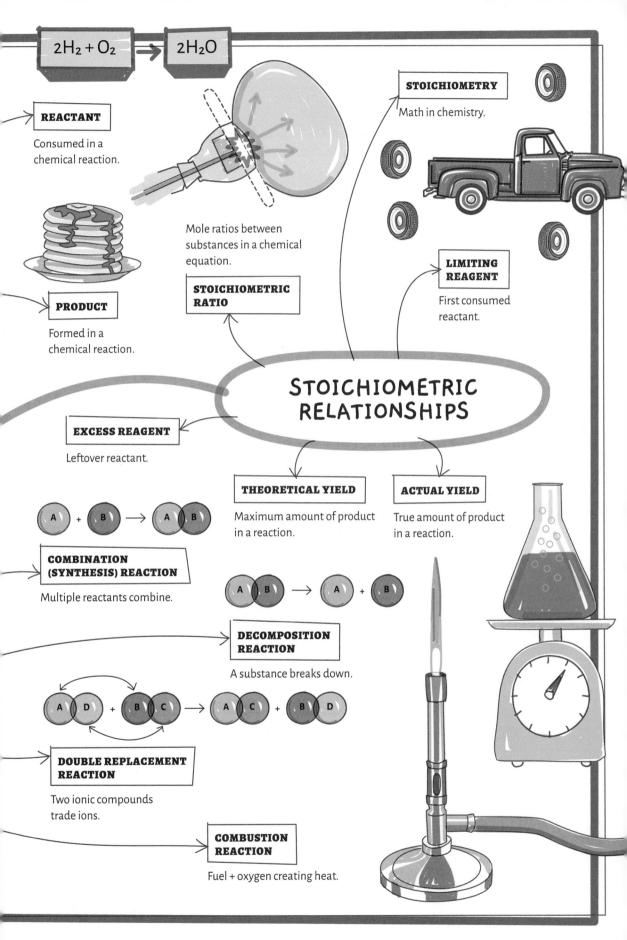

$2H_2 + O_2 \rightarrow 2H_2O$

REACTANT

Consumed in a chemical reaction.

PRODUCT

Formed in a chemical reaction.

STOICHIOMETRY

Math in chemistry.

Mole ratios between substances in a chemical equation.

STOICHIOMETRIC RATIO

LIMITING REAGENT

First consumed reactant.

STOICHIOMETRIC RELATIONSHIPS

EXCESS REAGENT

Leftover reactant.

THEORETICAL YIELD

Maximum amount of product in a reaction.

ACTUAL YIELD

True amount of product in a reaction.

COMBINATION (SYNTHESIS) REACTION

Multiple reactants combine.

DECOMPOSITION REACTION

A substance breaks down.

DOUBLE REPLACEMENT REACTION

Two ionic compounds trade ions.

COMBUSTION REACTION

Fuel + oxygen creating heat.

CHAPTER 9

SOLUTION CHEMISTRY

Solutions are homogeneous mixtures of two or more components. They play a crucial role in the biological, laboratory, and industrial applications of chemistry: the air we breathe, the liquids we drink, the blood in our veins, and the fluids in our bodies are all solutions. Aqueous solutions, which have water as one of the components, provide the medium for many important chemical and biological reactions that support life. For example, oxygen from our lungs combines chemically with hemoglobin in our red blood cells so it can be carried to body tissue. This life-supporting process would not be possible without the solution chemistry involved.

TYPES OF SOLUTION

A **solution** is a mixture of two or more components that form a single phase (liquid, solid, or gas) with a uniform molecular distribution throughout. The phase is determined by the component with the largest amount (the **solvent**); all other components are **solutes**, which are the substances that are dissolved in the solvent. The nature of the solvent and solute particles determines the characteristics of a solution, but those most commonly encountered in chemistry are aqueous solutions. Solution preparation is made possible by the **like dissolves like** principle.

Gaseous Solutions

In **gaseous solutions**, the solvent and solute molecules are both in the gas phase. Air, natural gas, and the gas mixture in a diver's breathing tank are gaseous solutions that we commonly encounter in daily life.

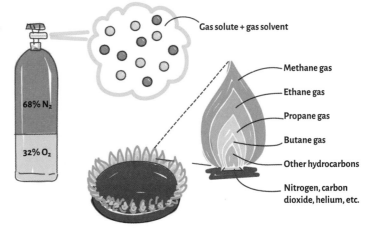

Gas solute + gas solvent

68% N_2

32% O_2

Methane gas
Ethane gas
Propane gas
Butane gas
Other hydrocarbons
Nitrogen, carbon dioxide, helium, etc.

Liquid Solutions

In a **liquid solution**, the solvent is in the liquid phase, but the solute can be in the solid, liquid, or gas phase. Liquid solutions of a solid solute are most often encountered in chemistry.

In carbonated water, gaseous carbon dioxide is dissolved in liquid water. This type of solution is not very stable, as the gaseous solute tends to escape from the solution. This is the reason why carbonated beverages are kept at cooler temperatures.

Saline solutions are often used in medicine. These are salt solutions in which water is the solvent and sodium chloride is the solute. For this type of solution the solute must be **soluble** in the solvent.

Rubbing alcohol is an example of a liquid solution where both solute and solvent are in liquid form before being mixed. The liquid components in the solution must be **miscible** with each other (they must be soluble in one another).

Solute

Solvent

Soda
Gas solute + liquid solvent

Saline
Solid solute + liquid solvent

Rubbing alcohol
Liquid solute + liquid solvent

Solid Solutions

Solutes and solvents in solid phase form **solid solutions**. In alloys, two solid metal components are mixed together to form a homogeneous mixture; steel, for example, is a mixture of iron (solvent) and carbon (solute).

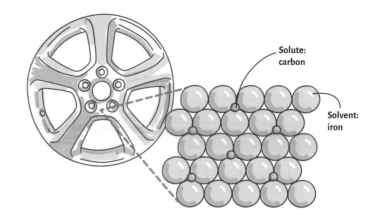

Solute: carbon

Solvent: iron

Concentrated and Dilute Solutions

Solutions can be classified based on the amount of solute they contain. A **concentrated solution** has a large amount of solute relative to the solvent, while a **dilute solution** is defined as one with a small amount of solute relative to the solvent.

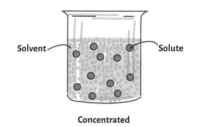

Solvent

Solute

Concentrated

Dilute

Saturated, Unsaturated, and Supersaturated Solutions

A solvent has a finite capacity in terms of how much solute it can dissolve at a given temperature. In an **unsaturated solution**, more solute can be dissolved in the solvent.

In a **saturated solution**, the solvent capacity has been reached. Any additional solute will settle at the bottom and will not be dissolved.

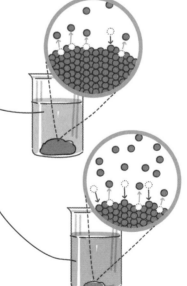

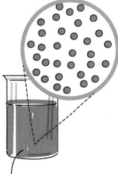

A **supersaturated solution** contains more solute than a saturated solution. The solvent capacity is exceeded, which makes this type of solution unstable.

Carbonated beverages are aqueous solutions in which gaseous carbon dioxide is dissolved under pressure. This is an example of a supersaturated solution. When a bottle containing such a beverage is opened, the pressure drops. This forces excess carbon dioxide to escape, as there is more carbon dioxide in the solution than the solvent can dissolve under the lower pressure conditions.

Solutions, Colloids, and Suspensions

The average particle size of the dissolved substance in a solution is less than 1 nanometer (1 nm = 1.0 × 10⁻⁹ m). Because of the small size of the solute particles, solutions appear uniform and they do not separate into layers.

The size of solute particles in **colloids** is between 1 and 100 nm, which is large enough to make colloids appear cloudy; milk, for example, is a colloid with tiny milk fat globs floating around the liquid. Colloids do not settle over time.

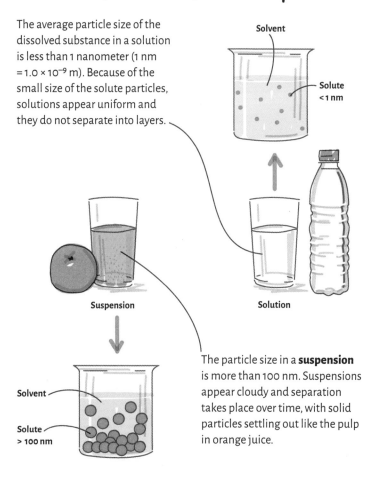

Solvent

Solute
<1 nm

Suspension

Solution

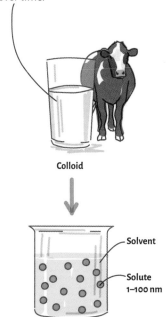

Colloid

The particle size in a **suspension** is more than 100 nm. Suspensions appear cloudy and separation takes place over time, with solid particles settling out like the pulp in orange juice.

Solvent

Solute
> 100 nm

Solvent

Solute
1–100 nm

Tyndall Effect

The scattering of a light beam by particles in colloids, suspensions, or air is known as the **Tyndall effect**. It is affected by the size of the particles present.

In a solution, the solute particles are too small to scatter light, so no Tyndall effect is observed; a light beam will pass right through the solution. However, as light passes through the larger solute particles in colloids and suspensions, it gets scattered by the solute particles and the beam becomes visible; the Tyndall effect is seen. This is similar to sunlight being scattered by dust particles in a room.

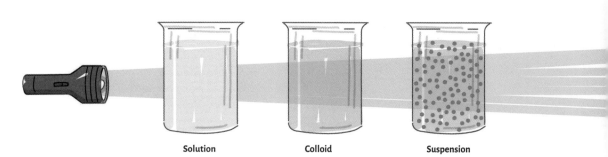

Solution

Colloid

Suspension

SOLUTION CONCENTRATION

The concentration of a solution is a measure of how much solute is dissolved in a given amount of solvent or solution and can quantitatively be expressed in different ways. The amount of solute in a solution is variable and depends on the purpose of preparing the solution. Many chemical reactions, for example, take place in solution, in which case knowing the exact amount of solute is important for stoichiometry.

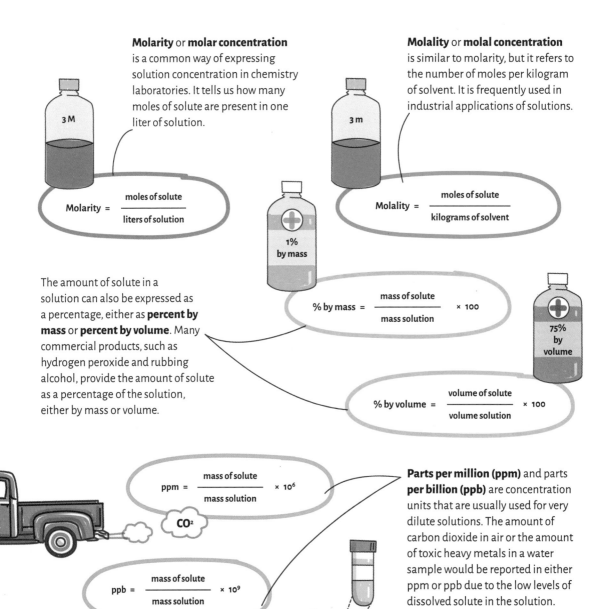

Molarity or **molar concentration** is a common way of expressing solution concentration in chemistry laboratories. It tells us how many moles of solute are present in one liter of solution.

3 M

$$\text{Molarity} = \frac{\text{moles of solute}}{\text{liters of solution}}$$

Molality or **molal concentration** is similar to molarity, but it refers to the number of moles per kilogram of solvent. It is frequently used in industrial applications of solutions.

3 m

$$\text{Molality} = \frac{\text{moles of solute}}{\text{kilograms of solvent}}$$

The amount of solute in a solution can also be expressed as a percentage, either as **percent by mass** or **percent by volume**. Many commercial products, such as hydrogen peroxide and rubbing alcohol, provide the amount of solute as a percentage of the solution, either by mass or volume.

1% by mass

$$\% \text{ by mass} = \frac{\text{mass of solute}}{\text{mass solution}} \times 100$$

75% by volume

$$\% \text{ by volume} = \frac{\text{volume of solute}}{\text{volume solution}} \times 100$$

$$\text{ppm} = \frac{\text{mass of solute}}{\text{mass solution}} \times 10^6$$

CO₂

$$\text{ppb} = \frac{\text{mass of solute}}{\text{mass solution}} \times 10^9$$

Parts per million (ppm) and parts **per billion (ppb)** are concentration units that are usually used for very dilute solutions. The amount of carbon dioxide in air or the amount of toxic heavy metals in a water sample would be reported in either ppm or ppb due to the low levels of dissolved solute in the solution.

PREPARATION OF SOLUTIONS

When preparing a solution the required mass or volume of solute is determined first, based on the desired concentration. The solution is then prepared by carefully mixing the solute and solvent substances.

Preparing a Stock Solution

A **stock solution** is a concentrated form of solution. Preparation begins by measuring out the necessary amount of solute. The amount could be in mass or volume depending upon the solution being prepared.

The solute is then transferred to a measuring container, such as a volumetric flask designed for the preparation of solutions with a specific volume. A line on the neck of the flask marks the volume of the container.

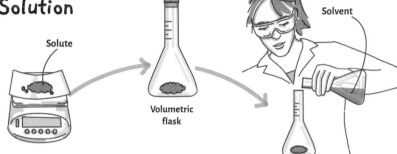

Solute

Volumetric flask

Solvent

The solute is dissolved in approximately half of the required amount of solvent needed for the solution. All solute particles must be dissolved for a uniform concentration.

More solvent is added to make the solution up to its final volume with the desired concentration.

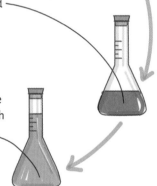

Stock solution

Dilution

Solutions can also be prepared by diluting a stock solution. **Serial dilution** techniques add varying amounts of solvent into a stock solution with a known concentration to achieve different dilution ratios.

Stock solution

Diluting a solution will increase its volume, as you are adding more solvent, and decrease its concentration. Chemists use the direct relationship between volume and molarity to dilute a stock solution.

$$V_{stock}M_{stock} = V_A M_A$$

10 mL stock solution + 90 mL solvent

1:10 dilution

$$V_A M_A = V_B M_B$$

10 mL solution A + 90 mL solvent

1:100 dilution

$$V_B M_B = V_C M_C$$

10 mL solution B + 90 mL solvent

1:1000 dilution

SOLUBILITY

The capacity of a substance to dissolve in a solvent is called its **solubility**. A solute will dissolve in a solvent up to its saturation point, which is the maximum amount of solute that a solvent can hold in solution. It is fundamentally dependent on the physical and chemical properties of the solvent and solute molecules ("like dissolves like"), as well as environmental conditions such as temperature, pressure, and other substances.

Solvation

If the attractive forces between the solute and solvent molecules are stronger than the forces keeping the solute particle together, the solute will dissolve in the solvent. In a process called **solvation**, solvent molecules surround the solute particles and pull them out into the solution.

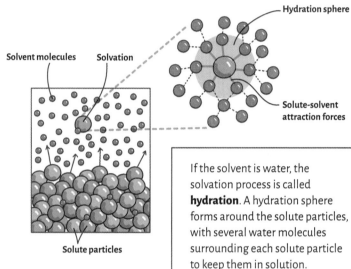

Solvent molecules   Solvation

Hydration sphere

Solute-solvent attraction forces

Solute particles

If the solvent is water, the solvation process is called **hydration**. A hydration sphere forms around the solute particles, with several water molecules surrounding each solute particle to keep them in solution.

Solution Equilibrium

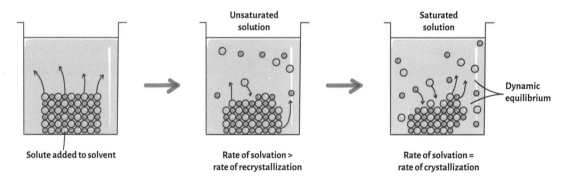

Unsaturated solution

Saturated solution

Dynamic equilibrium

Solute added to solvent

Rate of solvation > rate of recrystallization

Rate of solvation = rate of crystallization

As multiple solvent molecules are used to keep the solute particles in solution, there is a maximum number of solute particles that can be dissolved in any given volume of solvent. Beyond this limit the solute recrystallizes or precipitates out of solution.

The solvation process begins when the solute is first added to the solvent.

Solvation continues for as long as the solution remains unsaturated, which means the rate of solvation is greater than the rate of recrystallization or precipitation.

When solvent capacity is reached, the rate of solvation is equal to the rate of recrystallization. The number of solute particles dissolving per unit time is the same as the number of solute particles falling out of solution by recrystallization. The solution is said to have reached a **dynamic equilibrium**.

The Effect of Temperature

The solubility of solids and liquids generally increases with temperature. Molecules have greater kinetic energy at elevated temperatures, which improves solute-solvent interactions.

Solubility is increased due to the higher molecular speeds and the weakened intermolecular forces keeping the solute particles together. This is why a greater quantity of sugar can be dissolved in hot coffee compared to cold coffee.

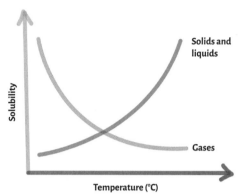

As temperature increases, the solubility of gases in liquids decreases. The faster molecular speeds at elevated temperatures enable gas molecules to escape the liquid state more effectively, which reduces their solubility. This is why carbonated beverages are best enjoyed chilled.

The Effect of Pressure

The solubility of solids and liquids is not affected by pressure, but gases have a higher solubility in liquids under increased pressure. When the pressure above them is increased, gas molecules are forced into the liquid phase, meaning that more gas molecules remain in solution.

Carbonated beverages are bottled under elevated pressure to dissolve carbon dioxide in water. When the bottle is opened, a hissing sound is heard because the pressure inside the bottle drops. As it does, dissolved carbon dioxide escapes the liquid phase due to reduced solubility under low-pressure conditions.

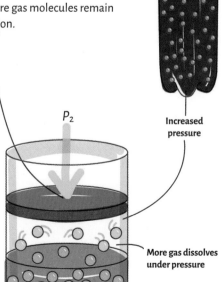

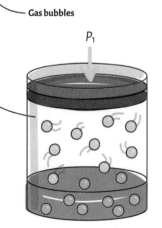

P_2

Increased pressure

Decreased pressure

P_1

Gas bubbles

More gas dissolves under pressure

CO_2 CO_2

SOLUBILITY RULES

The strong ion-dipole forces present in electrolyte solutions are responsible for the high solubility of ionic compounds. However, not all ionic compounds are soluble in water. The forces that keep some ionic compounds together are too strong to overcome when mixed with water, which makes such compounds insoluble. **Solubility rules** are a set of guidelines developed by chemists to quickly identify compounds in terms of their solubility tendencies.

Ionic compounds containing certain cations and anions, such as the group IA cations in the periodic table, are always soluble, with no insoluble exceptions.

Some combinations of ions form mostly soluble compounds, but there are a few exceptions that are insoluble. For example, most chloride (Cl^-) compounds are soluble, unless they are combined with Ag^+, Pb^{2+}, or Hg_2^{2+}, in which case they are not soluble.

Many hydroxide (OH^-) compounds are insoluble, but magnesium hydroxide exhibits slight solubility in water. It is used as the active ingredient in milk of magnesia to treat acid reflux.

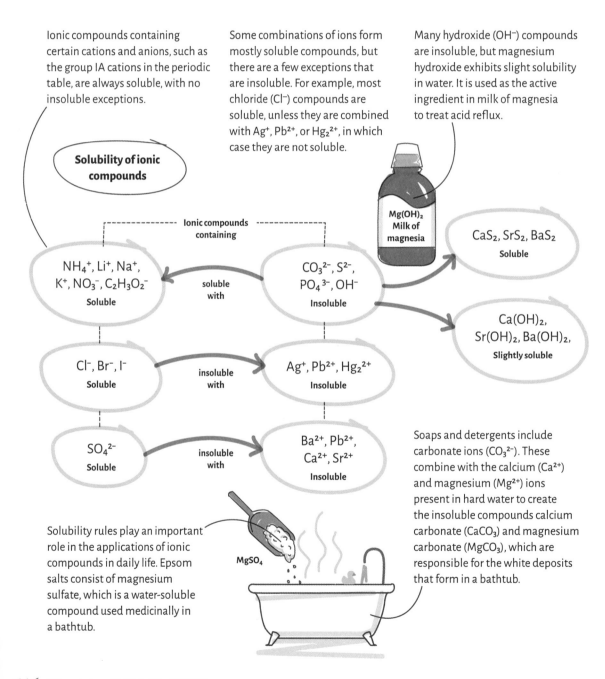

Solubility of ionic compounds

Ionic compounds containing

NH_4^+, Li^+, Na^+, K^+, NO_3^-, $C_2H_3O_2^-$
Soluble

soluble with

CO_3^{2-}, S^{2-}, PO_4^{3-}, OH^-
Insoluble

$Mg(OH)_2$ Milk of magnesia

CaS_2, SrS_2, BaS_2
Soluble

$Ca(OH)_2$, $Sr(OH)_2$, $Ba(OH)_2$,
Slightly soluble

Cl^-, Br^-, I^-
Soluble

insoluble with

Ag^+, Pb^{2+}, Hg_2^{2+}
Insoluble

SO_4^{2-}
Soluble

insoluble with

Ba^{2+}, Pb^{2+}, Ca^{2+}, Sr^{2+}
Insoluble

$MgSO_4$

Solubility rules play an important role in the applications of ionic compounds in daily life. Epsom salts consist of magnesium sulfate, which is a water-soluble compound used medicinally in a bathtub.

Soaps and detergents include carbonate ions (CO_3^{2-}). These combine with the calcium (Ca^{2+}) and magnesium (Mg^{2+}) ions present in hard water to create the insoluble compounds calcium carbonate ($CaCO_3$) and magnesium carbonate ($MgCO_3$), which are responsible for the white deposits that form in a bathtub.

Precipitation Reactions

A **precipitation reaction** is defined as the formation of an insoluble compound when solutions of two water-soluble compounds are mixed. For example, potassium iodide (KI) and lead (II) nitrate $(Pb(NO_3)_2)$ are two soluble compounds, but when their aqueous solutions are combined, the water-insoluble compound lead (II) iodide (PbI_2) forms as a yellow precipitate.

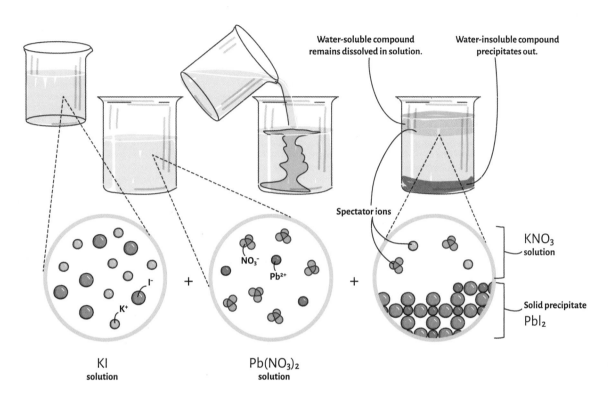

Water-soluble compound remains dissolved in solution.

Water-insoluble compound precipitates out.

Spectator ions

NO_3^-

Pb^{2+}

I^-

K^+

KNO$_3$ solution

Solid precipitate PbI_2

KI solution

$Pb(NO_3)_2$ solution

The **molecular equation** shows the double displacement reaction that takes place when you mix the two aqueous solutions.

$$2KI\ (aq) + Pb(NO_3)_2\ (aq) \rightarrow PbI_2\ (s) + 2KNO_3\ (aq)$$

$$2K^+\ (aq) + 2I^-\ (aq) + Pb^{2+}\ (aq) + 2NO_3^-\ (aq) \rightarrow PbI_2\ (s) + 2K^+\ (aq) + 2NO_3^-\ (aq)$$

The **total ionic equation** for the reaction is a better representation of the actual particles present in the solutions. All soluble compounds are shown as solvated ions, while the precipitate is written in the compound form, as it does not dissolve into solvated ions.

$$Pb^{2+}\ (aq) + 2I^-\ (aq) \rightarrow PbI_2\ (s)$$

The **net ionic equation** only shows the ions participating in the formation of the precipitate. All of the other ions that are not involved in the formation of the solid product are **spectator ions**, so they are not shown. In the reaction shown here, K^+ and NO_3^- are spectator ions.

COLLIGATIVE PROPERTIES OF SOLUTIONS

Taken from the Latin *colligatus*, which means "bound together," **colligative properties** are based on the number of solute particles in a solution, rather than their chemical identity. These properties demonstrate the differences in the molecular environment between the pure solvent and the solution.

Vapor Pressure Lowering

Liquid molecules at the surface of a liquid escape into the gas phase, which will create pressure above the liquid if it is in a closed container. The **vapor pressure** of a pure liquid is defined as the pressure exerted by a vapor in equilibrium with its liquid state. How much pressure the vapor can exert at room temperature is closely related to the strength of its intermolecular forces in the liquid phase—liquids with a weaker intermolecular force have a higher vapor pressure.

The vapor pressure of a pure solvent is lowered when a nonvolatile solute is dissolved in it, because the presence of solute molecules disrupts the molecular structure of the liquid. As solute-solvent attractions are stronger than solvent-solvent attractions, fewer liquid molecules can escape into the gas phase, reducing the vapor pressure of the solution to below that of the pure solvent.

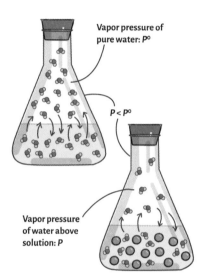

Vapor pressure of pure water: P^o

$P < P^o$

Vapor pressure of water above solution: P

Freezing Point Depression

Freezing point depression refers to the observation that the freezing temperature of a solution is lower than that of the pure solvent.

The ions present in a solution of salt water interrupt the hexagonal structure of ice formation that is found in pure water, causing salt water to freeze at a temperature lower than 0°C. The more ions that are present in the water, the lower the freezing point becomes.

Therefore, electrolytes with a high number of ions depress the freezing point more than nonelectrolytes for aqueous solutions. This is why icy roads are treated with salt in the winter months.

m = solution molality

K_f = solvent constant

ΔT_f = change in freezing point

$$\Delta T_f = m \times K_f$$

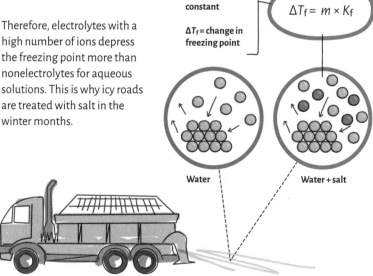

Water

Water + salt

Boiling Point Elevation

Because solute-solvent attractions are stronger than solvent-solvent attractions, more energy is needed to boil a solution than is required to boil the pure solvent. Boiling starts when the vapor pressure above the liquid becomes identical to the atmospheric pressure. However, as solutions have lower vapor pressure than pure solvents, more molecules are required in the gas phase to match the atmospheric pressure so boiling can begin. This process requires higher temperatures. **Boiling point elevation** is the observation that a solution has a higher boiling temperature than the pure solvent.

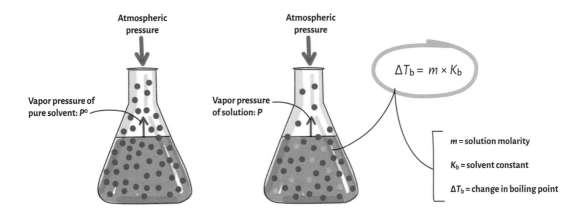

Atmospheric pressure

Atmospheric pressure

Vapor pressure of pure solvent: P^o

Vapor pressure of solution: P

$$\Delta T_b = m \times K_b$$

m = solution molarity

K_b = solvent constant

ΔT_b = change in boiling point

Osmotic Pressure

Osmosis is the process of solvent movement from a dilute solution to a concentrated solution. Concentrated saline solutions are often used to treat constipation because the solution draws water from the surrounding tissue as it moves through the intestines, alleviating the medical condition.

The intestinal wall acts like an osmosis cell in which a semipermeable membrane separates two solutions, one dilute and one concentrated. The membrane allows the solvent molecules to pass through, but not the solute particles. Over time, the liquid level on the dilute side drops, as the solvent moves into the concentrated cell. The difference in the liquid levels is known as the **osmotic pressure**.

External pressure can be applied to the concentrated cell to reverse the flow of the solvent, so it passes from the concentrated cell to the dilute side. This process is called **reverse osmosis** and forms the basis of producing drinking water from seawater.

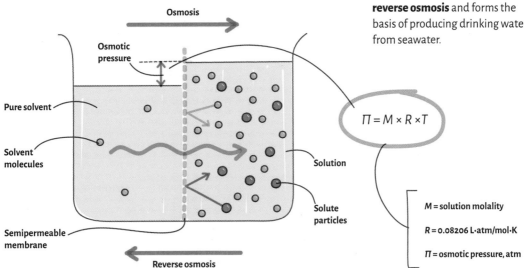

Osmosis

Osmotic pressure

Pure solvent

Solvent molecules

Semipermeable membrane

Reverse osmosis

Solution

Solute particles

$$\Pi = M \times R \times T$$

M = solution molality

R = 0.08206 L·atm/mol·K

Π = osmotic pressure, atm

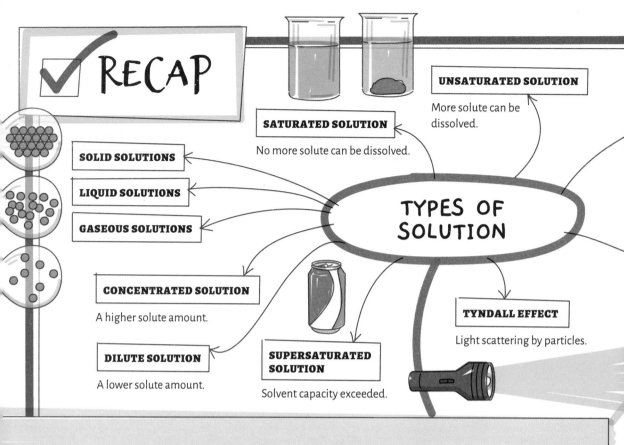

✓ RECAP

SATURATED SOLUTION
No more solute can be dissolved.

UNSATURATED SOLUTION
More solute can be dissolved.

SOLID SOLUTIONS

LIQUID SOLUTIONS

GASEOUS SOLUTIONS

TYPES OF SOLUTION

CONCENTRATED SOLUTION
A higher solute amount.

DILUTE SOLUTION
A lower solute amount.

SUPERSATURATED SOLUTION
Solvent capacity exceeded.

TYNDALL EFFECT
Light scattering by particles.

SOLUTION CHEMISTRY

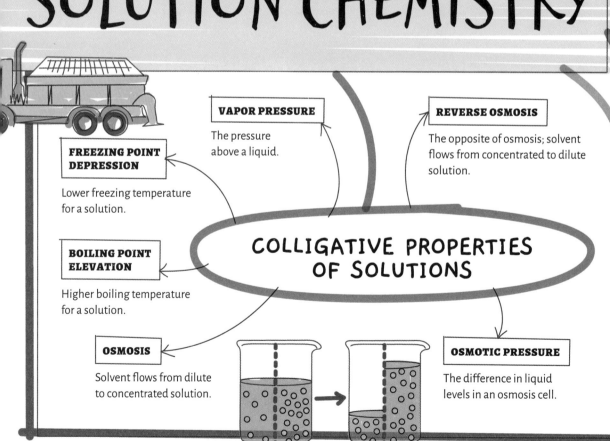

VAPOR PRESSURE
The pressure above a liquid.

REVERSE OSMOSIS
The opposite of osmosis; solvent flows from concentrated to dilute solution.

FREEZING POINT DEPRESSION
Lower freezing temperature for a solution.

BOILING POINT ELEVATION
Higher boiling temperature for a solution.

COLLIGATIVE PROPERTIES OF SOLUTIONS

OSMOSIS
Solvent flows from dilute to concentrated solution.

OSMOTIC PRESSURE
The difference in liquid levels in an osmosis cell.

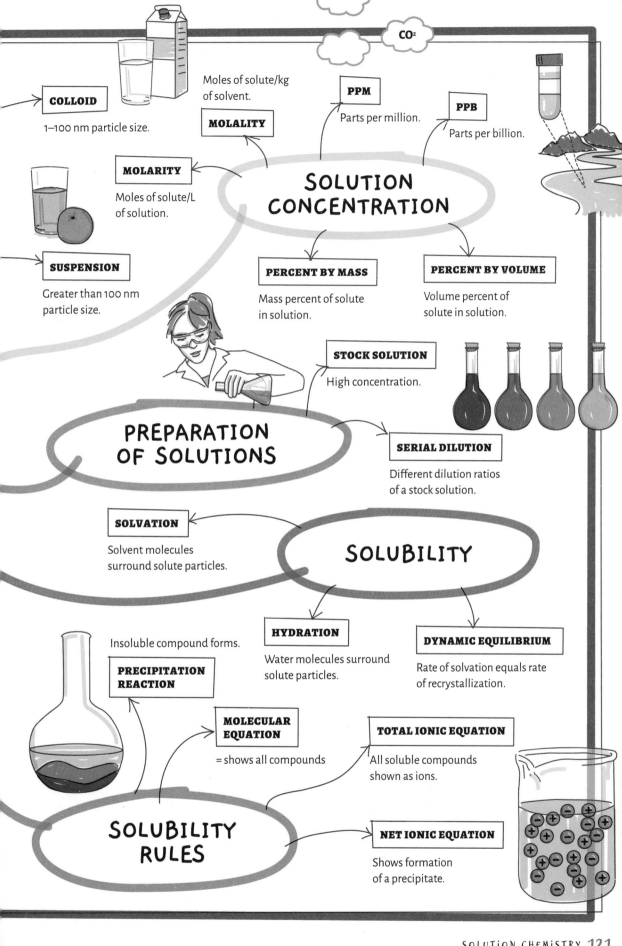

COLLOID

1–100 nm particle size.

Moles of solute/kg of solvent.

MOLALITY

PPM

Parts per million.

PPB

Parts per billion.

MOLARITY

Moles of solute/L of solution.

SOLUTION CONCENTRATION

SUSPENSION

Greater than 100 nm particle size.

PERCENT BY MASS

Mass percent of solute in solution.

PERCENT BY VOLUME

Volume percent of solute in solution.

STOCK SOLUTION

High concentration.

PREPARATION OF SOLUTIONS

SERIAL DILUTION

Different dilution ratios of a stock solution.

SOLVATION

Solvent molecules surround solute particles.

SOLUBILITY

Insoluble compound forms.

HYDRATION

Water molecules surround solute particles.

DYNAMIC EQUILIBRIUM

Rate of solvation equals rate of recrystallization.

PRECIPITATION REACTION

MOLECULAR EQUATION

= shows all compounds

TOTAL IONIC EQUATION

All soluble compounds shown as ions.

SOLUBILITY RULES

NET IONIC EQUATION

Shows formation of a precipitate.

CHAPTER 10

GASES

Gas is one of the three fundamental physical states of matter. It is characterized by a large separation between its particles, compared to the liquid and solid phases, and doesn't have a fixed shape or volume. Most gases are invisible to the human eye, but a molecular understanding of how they behave under different conditions plays an important part in the work that many chemists and other scientists undertake. A pure gas may contain individual atoms (He, Ne), molecules of the same type of atom (H_2, N_2, O_2), or molecules of different atoms (CO_2, SO_2), but gases often exist as a mixture of two or more pure gases. For example, the gas we encounter most often is Earth's atmosphere, which is a gigantic "sea" of gas mixture.

KINETIC MOLECULAR THEORY

The particle nature of matter is used to explain the observed properties of gases in **kinetic molecular theory**. The theory is based on the notion that gas particles are in ceaseless random motion in their container, which gives rise to various physical properties, such as temperature, pressure, and volume. Although this is an unrealistic picture of how real gases behave, the fundamental ideas developed from it apply to all gases.

Basic Postulates

★ There are no attractive or repulsive forces between gas particles. They essentially ignore each other.

★ Gas particles, atoms, and molecules are very small in size compared to the distance that separates them. Gas particles have negligible volume.

★ Gas particles are in ceaseless random motion in their container. This is called **Brownian motion** and it results in collisions with other particles and the container walls.

★ No energy is lost when gas particles collide with each other. This type of particle collision is known as **elastic collision**, in which particles may exchange energy with each other, but the overall energy is conserved.

★ Regardless of identity, all gases have the same average kinetic energy at the same temperature. Their average kinetic energy is proportional to their temperature in Kelvin.

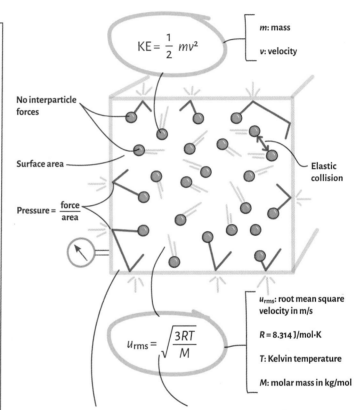

$$KE = \frac{1}{2} mv^2$$

m: mass
v: velocity

No interparticle forces

Surface area

$$Pressure = \frac{force}{area}$$

Elastic collision

u_{rms}: root mean square velocity in m/s

$R = 8.314$ J/mol·K

T: Kelvin temperature

M: molar mass in kg/mol

$$u_{rms} = \sqrt{\frac{3RT}{M}}$$

The collision of gas particles with the walls of the container creates **pressure**, which is the total force exerted by the gas particles per unit area of the container walls. Pressure can be measured easily by connecting a pressure gauge to one of the walls of the container.

The average speed of gas molecules in a container is given as the **root mean square velocity**. This takes temperature and molecular weight into account, which are two primary factors in determining how fast a gas molecule can move. Consequently, it shows that speed is directly proportional to Kelvin temperature and inversely proportional to molecular weight.

GAS LAWS

There are several gas laws that describe the behavior of a gas under varying conditions of temperature (*T*), pressure (*P*), volume (*V*), and moles of gas (*n*). As these four basic properties are interrelated, when any one of them changes, the others are also affected. The simple gas laws explain the relationship between one pair of these properties at a time, while keeping the other two unchanged; combined, the simple gas laws create the **ideal gas law**.

The popping we experience in our ears while flying on an airplane is also a consequence of Boyle's law. At high altitudes, the air pressure is much lower than it is at sea level. The eardrum "pops" out to increase the air volume inside the ear because of the low pressure experienced during a flight. To minimize the discomfort felt by passengers, the cabin on an airplane is pressurized.

Boyle's Law

Boyle's law involves the relationship between volume and pressure. The amount of gas and the temperature remain constant.

Kinetic molecular theory indicates that the size of gas particles is negligibly small compared to the distance between them. Therefore, if a certain amount of gas is confined to a smaller volume at the same temperature, the particles will get closer to each other. This increases the frequency of interparticle and particle-wall collisions, leading to an increase in pressure.

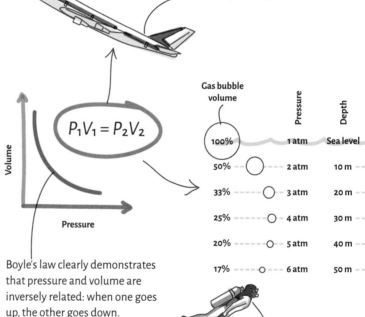

$$P_1V_1 = P_2V_2$$

Boyle's law clearly demonstrates that pressure and volume are inversely related: when one goes up, the other goes down.

Boyle's law has many applications in the world. For example, a diver needs to be careful when ascending and descending in water. Divers feel increased pressure as they descend, due to the larger amount of water above them. This increased pressure will reduce the volume of air in the diver's lungs. The opposite happens when a diver ascends to the surface. The change in air volume in a diver's body can be lethal, which is why diving is highly regulated and requires special equipment.

Charles's Law

Charles's law explains the relationship between temperature and volume for a gas; the pressure and the amount (in moles) remain constant. The average kinetic energy of gas particles rises with increasing temperature. To keep the pressure constant, the volume of the gas also increases, due to higher molecular speeds.

According to Charles's law, temperature and volume are directly proportional under constant pressure conditions for a certain amount of gas.

As the temperature drops, the volume decreases and eventually reaches zero. This theoretical point corresponds to the **absolute zero temperature** with a value of 0 Kelvin.

A hot air balloon can float because the air inside it is heated up to increase its volume and reduce its density.

$$\frac{V_1}{T_1} = \frac{V_2}{T_2}$$

Temperature (K)

Volume

Absolute zero temperature = -273.15°C = 0 K

V_1, T_1

Balloon in ice water

V_2, T_2

Balloon in boiling water

He

$$\frac{V_1}{n_1} = \frac{V_2}{n_2}$$

Volume

Moles

Avogadro's Law

According to Avogadro's law, the volume of a gas is directly proportional to the amount of moles of available gas at a constant temperature and pressure. We experience this when inflating a balloon: the more gas in the balloon, the larger its size becomes.

An important consequence of Avogadro's law is that one mole of any gas, regardless of its identity, occupies 22.4 liters of volume under standard temperature and pressure conditions (0°C / 1 atm).

1 mol He	1 mol NH_3	1 mol O_2
$V = 22.4$ L	$V = 22.4$ L	$V = 22.4$ L
$P = 1$ atm	$P = 1$ atm	$P = 1$ atm
$T = 273.15$ K	$T = 273.15$ K	$T = 273.15$ K

Gay-Lussac's Law

Gay-Lussac's law states that the pressure of a given amount of gas varies directly with its Kelvin temperature under constant volume conditions and a fixed amount of moles of gas.

Food cooks faster in a pressure cooker because the steam inside is much hotter in the fixed volume of the pot than normal steam exposed to atmospheric pressure. Care must be taken while using a pressure cooker,

though, because the increased temperature causes the pressure to rise significantly. A relief valve allows some of the steam to escape, preventing the pressure from building up too much.

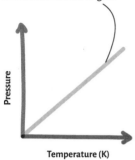

$$\frac{P_1}{T_1} = \frac{P_2}{T_2}$$

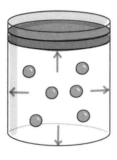

P_1, T_1

P_2, T_2

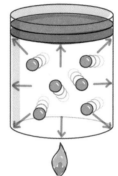

Ideal Gas Law

The four simple gas laws combine into a single law known as the **ideal gas law**. The ideal gas law encompasses all of the relationships between the properties of a gas.

$$P_1V_1 = P_2V_2$$

$$\frac{V_1}{n_1} = \frac{V_2}{n_2}$$

$$PV = nRT$$

$$\frac{V_1}{T_1} = \frac{V_2}{T_2}$$

$$\frac{P_1}{T_1} = \frac{P_2}{T_2}$$

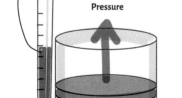

R is the universal gas constant, with a fixed numerical value.

$$R = 0.08206 \ \frac{\text{L·atm}}{\text{mol·K}}$$

The ideal gas law equation shows that when one of the properties of a gas changes, the others will change as well. Knowing the pressure, temperature, volume,

and mole amount of a gas defines the complete state of the gas. Therefore, the ideal gas law equation is known as an **equation of state**.

Combined Gas Law

The **combined gas law** describes the relationship between pressure, temperature, and volume of a certain fixed amount of gas in an enclosed container. It therefore provides information about what happens to the properties of a gas when the environmental conditions change.

A weather balloon is used to collect atmospheric data including air pressure, temperature, and wind speed. The balloon is filled with helium, which is less dense than air, and carries measuring instruments to different altitudes. The air temperature and pressure changes at different altitudes cause the balloon to change size.

$$\frac{P_1 V_1}{T_1} = \frac{P_2 V_2}{T_2}$$

Balloon bursts

Parachute

Instrument gathers data

Air pressure drops significantly with altitude. This causes the helium gas in a rising weather balloon to expand, as it compensates for the decrease in pressure. The balloon continues to increase in size as it rises until, at approximately 27,000 meters, it bursts. The measuring device returns to the ground by parachute.

$$\text{Density} = \frac{PM}{RT}$$

Gas Density

The density of a gas can be determined from the ideal gas law and is directly proportional to its molar mass (M): the heavier the gas, the higher its density. Air with a molar mass of 29 g/mol has a density of 1.18 g/L at ambient atmospheric conditions, while helium with a molar mass of 4 g/mol exhibits a density of 0.164 g/L. This difference in densities is the reason why a helium-filled weather balloon can lift measuring devices into the atmosphere. The size of the balloon is determined based on its intended payload.

GAS MIXTURES

Many of the gases we encounter in daily life are not pure. Air, for example, is a mixture of oxygen, nitrogen, carbon dioxide, argon, and trace amounts of some other gases as well. Each component of a gas mixture is treated independently because it has a negligible particle size and there are no interactions between the components according to the kinetic molecular theory of gases. Consequently, the volume and the temperature are identical for all component gases, but their individual pressures may be different depending upon their mole amounts.

Dalton's Law

The pressure of a component in a gas mixture is called its **partial pressure**. **Dalton's law** states that the total pressure of a gas mixture is the sum of the partial pressures of all its components.

The partial pressure of a component in a gas mixture is its **mole fraction** (X) multiplied by the total pressure. The mole fraction is the ratio of the number of moles of one component to the total number of moles in the mixture.

Gas A

P_A
n_A

Gas B

P_B
n_B

Gas C

P_C
n_C

$$T_A = T_B = T_C \qquad V_A = V_B = V_C$$

The ideal gas law applies to pure gases, as well as their mixtures.

$$X_A = n_A/n_{total} \qquad P_{total}V = n_{total}RT$$

$$P_{total} = P_A + P_B + P_C$$

$$P_A = X_A \cdot P_{total} \qquad P_B = X_B \cdot P_{total} \qquad P_C = X_C \cdot P_{total}$$

There are many applications of Dalton's law in daily life. The air around us, for example, is a gas mixture that always contains approximately 21 percent oxygen by volume. At sea level, where the atmospheric pressure is 1 atm, the partial pressure of oxygen is 0.20 atm. However, on top of a mountain the atmospheric pressure drops. The partial pressure of oxygen can be as low as 0.066 atm, which is not enough for respiratory comfort, and explains why mountain climbers can experience **hypoxia** (low oxygen levels), which causes headaches, dizziness, and a shortness of breath. At extremely high elevations, such as the peak of Mount Everest, the oxygen level in the air is so low that it can cause unconsciousness or even death. This is why most climbers attempting the summit will use an oxygen tank.

High altitude

$P_{oxygen} = 0.066$ atm

Atmospheric pressure

$P_{oxygen} = 0.20$ atm

Sea level

10000 m
8000 m
6000 m
4000 m
2000 m
0

CHEMICAL REACTIONS

Many chemical reactions take place in the gaseous phase, or involve at least one gaseous reactant or product. The stoichiometric amount of gases involved in a chemical reaction is easier to measure in volume rather than mass; the ideal gas law equation provides the necessary mathematical expression to convert between moles and volume.

The Law of Combining Gas Volumes

If all of the gases in a chemical reaction are at the same temperature and pressure, then the stoichiometric coefficients represent both the gas volumes and the mole numbers. This is known as the **law of combining gas volumes**.

In the production of ammonia (NH$_3$), the 1:3:2 stoichiometric mole ratios also represent the volume ratios. Therefore, if 1 L of nitrogen gas is combined with 3 L of hydrogen gas, 2 L of ammonia gas will be produced.

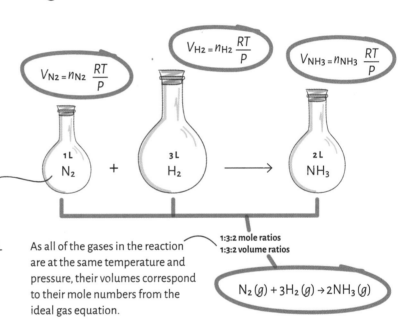

$$V_{N2} = n_{N2}\frac{RT}{P}$$

$$V_{H2} = n_{H2}\frac{RT}{P}$$

$$V_{NH3} = n_{NH3}\frac{RT}{P}$$

1 L
N$_2$ + 3 L H$_2$ → 2 L NH$_3$

1:3:2 mole ratios
1:3:2 volume ratios

$$N_2\,(g) + 3H_2\,(g) \rightarrow 2NH_3\,(g)$$

As all of the gases in the reaction are at the same temperature and pressure, their volumes correspond to their mole numbers from the ideal gas equation.

Gas Collection Over Water

Chemists frequently carry out chemical reactions to produce a gaseous product of interest—zinc (Zn), for example, reacts with hydrochloric acid (HCl) to form hydrogen gas (H$_2$). One of the most convenient methods employed in a chemistry laboratory to collect the gas that is generated is water displacement, which forms a gas mixture of water vapor and hydrogen gas. Dalton's law is applied to find the moles of hydrogen gas in stoichiometry calculations.

$$P_{total} = P_{H_2} + P_{H_2O}$$

H$_2$O

H$_2$

Zn

HCl

$$Zn\,(s) + 2HCl\,(aq) \rightarrow H_2\,(g) + ZnCl_2\,(aq)$$

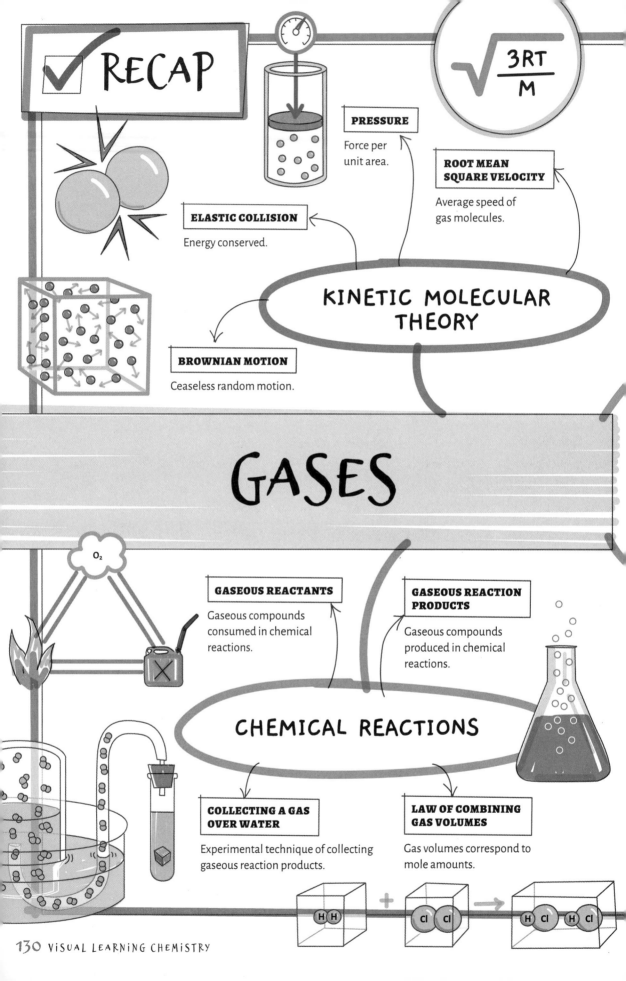

RECAP

$$\sqrt{\frac{3RT}{M}}$$

PRESSURE

Force per unit area.

ROOT MEAN SQUARE VELOCITY

Average speed of gas molecules.

ELASTIC COLLISION

Energy conserved.

KINETIC MOLECULAR THEORY

BROWNIAN MOTION

Ceaseless random motion.

GASES

O_2

GASEOUS REACTANTS

Gaseous compounds consumed in chemical reactions.

GASEOUS REACTION PRODUCTS

Gaseous compounds produced in chemical reactions.

CHEMICAL REACTIONS

COLLECTING A GAS OVER WATER

Experimental technique of collecting gaseous reaction products.

LAW OF COMBINING GAS VOLUMES

Gas volumes correspond to mole amounts.

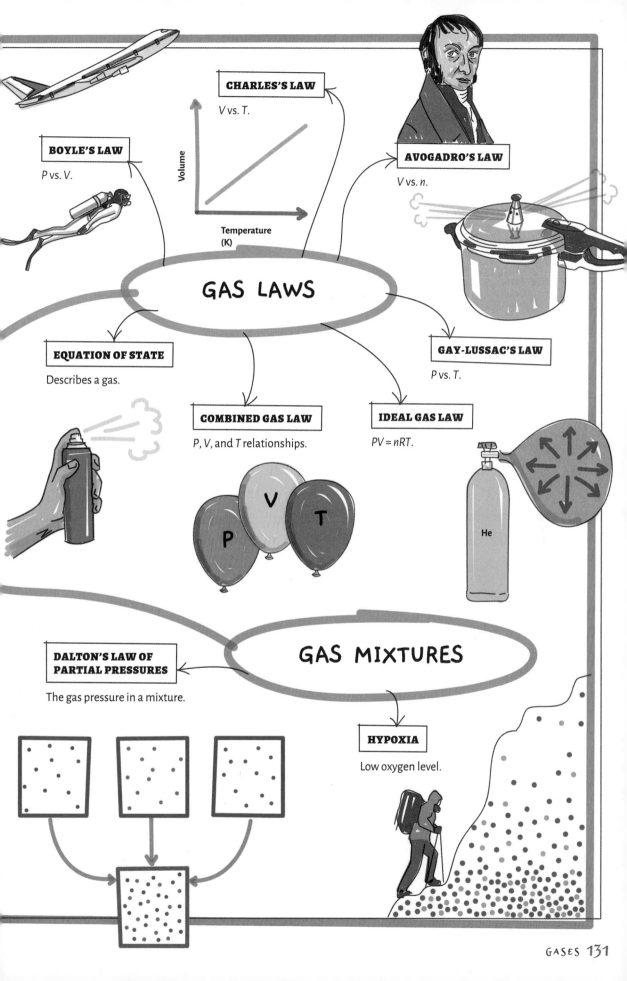

CHARLES'S LAW

V vs. *T*.

BOYLE'S LAW

P vs. *V*.

Volume

Temperature (K)

AVOGADRO'S LAW

V vs. *n*.

GAS LAWS

EQUATION OF STATE

Describes a gas.

GAY-LUSSAC'S LAW

P vs. *T*.

COMBINED GAS LAW

P, *V*, and *T* relationships.

IDEAL GAS LAW

$PV = nRT$.

P V T

He

GAS MIXTURES

DALTON'S LAW OF PARTIAL PRESSURES

The gas pressure in a mixture.

HYPOXIA

Low oxygen level.

CHAPTER 11

CHEMICAL EQUILIBRIUM

The conventional way of writing chemical equations suggests that a reaction proceeds in one direction, with the reactants being converted into products until all of the reactant molecules are consumed. In reality, though, most reactions are reversible. When reactants are mixed together, the reaction begins and proceeds in the forward direction, creating products. However, as long as product molecules are present in the reaction mixture, the reverse reaction can also take place, whereby product molecules break down into reactants.

DYNAMIC EQUILIBRIUM

When the rate of forward and reverse reactions is equal, the reaction has reached **equilibrium**. At this point no further changes will occur in the amounts of reactants and products. However, this does not mean that the reaction has come to an end—the forward and reverse reactions are still taking place, but they are happening at the same speed. This is defined as **dynamic equilibrium**.

The Formation of Dynamic Equilibrium

Dynamic equilibrium is established for the water level in a tank if identical amounts of water are entering and leaving. Incoming water is the "forward reaction" while outflowing water represents the "reverse reaction."

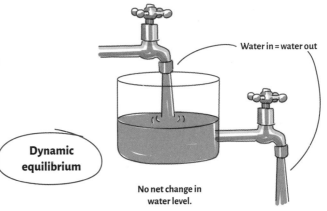

Water in = water out

Dynamic equilibrium

No net change in water level.

At the start of the decomposition reaction of dinitrogen tetroxide (N_2O_4), only N_2O_4 molecules are present, but as soon as the reaction begins, nitrogen dioxide (NO_2) starts to form. As long as the rate of forward reaction is higher than the rate of reverse reaction, the amount of N_2O_4 will decrease, while the concentration of NO_2 will increase.

At dynamic equilibrium, there is no net change in the N_2O_4 and NO_2 concentrations in the reaction mixture, as the forward and reverse reaction rates are now equal. A double-headed arrow is used to designate equilibrium in a chemical equation.

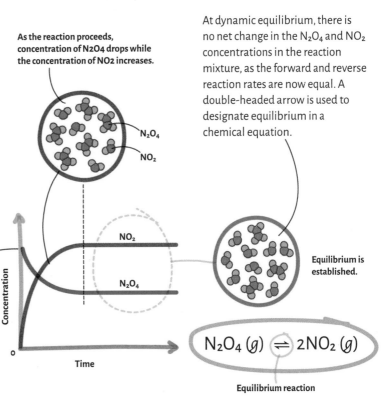

As the reaction proceeds, concentration of N2O4 drops while the concentration of NO2 increases.

N_2O_4

NO_2

Equilibrium is established.

Concentration

NO_2

N_2O_4

Time

At time zero: only N_2O_4 is present.

$$N_2O_4 \,(g) \rightleftharpoons 2NO_2 \,(g)$$

Equilibrium reaction

EQUILIBRIUM CONSTANT

Once equilibrium is established in a chemical reaction, there will be no net change in the concentrations of the reactants and products unless there are changes in the reaction conditions, such as a change in temperature (although the concentrations will not necessarily be equal). The relative concentrations of reactants and products at equilibrium are expressed in a quantity called the **equilibrium constant (K)**.

The Law of Mass Action

The **law of mass action** states that for a reversible reaction at equilibrium, and at a constant temperature, a certain ratio of product and reactant

concentrations will have a constant value. This law provides a mathematical definition for the equilibrium constant (K).

For a generic reaction, an expression for the equilibrium constant can be written as the ratio of the concentrations in molarity of reactants (A and B) and products (C and D), raised to the power of their stoichiometric coefficients. K is a constant, with no units.

Equilibrium constant in terms of concentration

Products

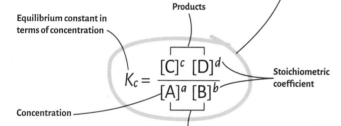

$$K_c = \frac{[C]^c \, [D]^d}{[A]^a \, [B]^b}$$

Stoichiometric coefficient

Concentration

Reactants

$$aA + bB \Leftrightarrow cC + dD$$

$$K_c = \frac{[NO_2]^2}{[N_2O_4]}$$

$$K_p = \frac{p^2_{NO_2}}{p_{N_2O_4}}$$

$$K_p = K_c(RT)^{\Delta n}$$

The equilibrium constant can be expressed in terms of **concentrations (K_c)** or—when the reactants and products are in the gaseous state—as **partial pressures (K_p)**. For the decomposition reaction of N_2O_4 into NO_2 the K_p expression can be used, as both molecules are gases at room temperature.

The ideal gas law provides the mathematical relationship between K_p and K_c for a chemical reaction at equilibrium in the gas phase. In this equation, R is the universal gas constant, T is temperature in Kelvin, and Δn is the difference between the sum of stoichiometric coefficients of gaseous products and reactants (for the decomposition reaction of N_2O_4, $\Delta n = 1$).

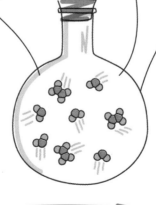

$$N_2O_4 \, (g) \; \rightleftharpoons \; 2NO_2 \, (g)$$

Heterogeneous Equilibrium

Pure liquids and solids that coexist with a gaseous or an aqueous phase are excluded from the K expression for heterogeneous reactions. As long as there is some pure solid or liquid present in a reaction, their concentration does not change, so they do not appear in the equilibrium constant.

Solid carbon does not appear in the K expression because its concentration is not affected.

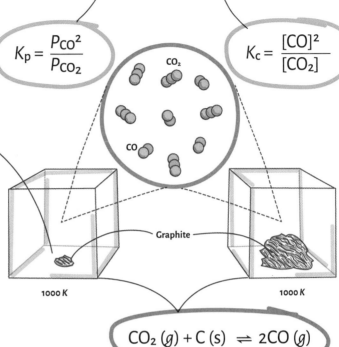

$$K_p = \frac{P_{CO}^2}{P_{CO_2}}$$

$$K_c = \frac{[CO]^2}{[CO_2]}$$

Carbon dioxide (CO_2) gas can react with graphite (carbon [C]) at high temperatures to form gaseous carbon monoxide (CO). The equilibrium between CO_2 and CO in the gas phase is not affected by the amount of solid carbon, as long as it is present in the reaction mixture.

Graphite

1000 K   1000 K

$$CO_2\ (g) + C\ (s) \rightleftharpoons 2CO\ (g)$$

The Significance of K

K provides the concentration ratio of products to reactants at equilibrium conditions. The value of K essentially shows how much product can be made under a set of reaction conditions.

A large K value indicates that the concentration of products is much higher than that of reactants at equilibrium. Consequently, equilibrium lies toward the products.

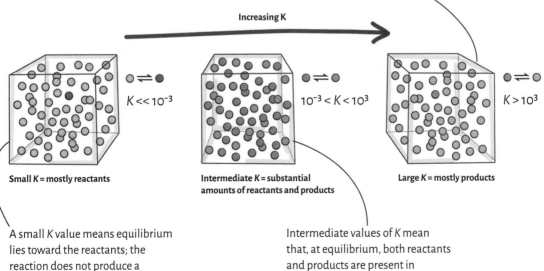

Increasing K

$K \ll 10^{-3}$

$10^{-3} < K < 10^3$

$K > 10^3$

Small K = mostly reactants

Intermediate K = substantial amounts of reactants and products

Large K = mostly products

A small K value means equilibrium lies toward the reactants; the reaction does not produce a high concentration of products.

Intermediate values of K mean that, at equilibrium, both reactants and products are present in appreciable amounts.

REACTION QUOTIENT

nitially, when there are only reactant molecules present, a chemical reaction inevitably proceeds in the forward direction to make the products. For reversible reactions, the opposite is also true: if only product molecules are present at the outset, the reaction proceeds in the reverse direction toward the reactants. Under nonequilibrium conditions, where both reactant and product molecules are present, the progress of a chemical reaction and the direction in which it is moving are determined by the **reaction quotient**.

K Versus Q

The reaction quotient is defined exactly the same way as K, but the reaction does not have to be at equilibrium. The concentrations used in the Q expression do not correspond to equilibrium concentrations.

Q tells us where the reaction lies at any given moment in relation to equilibrium. If it is not already at equilibrium, the numerical value of Q compared to K indicates whether the reaction is moving in the forward or reverse direction.

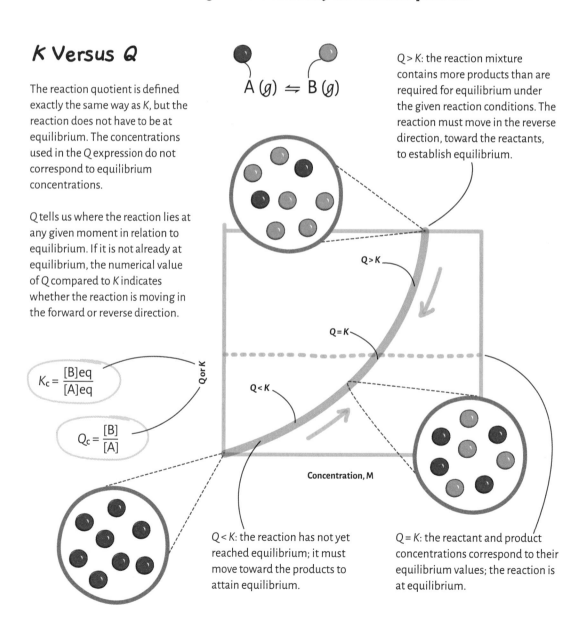

$$A\ (g) \rightleftharpoons B\ (g)$$

$$K_c = \frac{[B]eq}{[A]eq}$$

$$Q_c = \frac{[B]}{[A]}$$

$Q > K$

$Q = K$

$Q < K$

Q or K

Concentration, M

$Q > K$: the reaction mixture contains more products than are required for equilibrium under the given reaction conditions. The reaction must move in the reverse direction, toward the reactants, to establish equilibrium.

$Q < K$: the reaction has not yet reached equilibrium; it must move toward the products to attain equilibrium.

$Q = K$: the reactant and product concentrations correspond to their equilibrium values; the reaction is at equilibrium.

CHANGING EQUILIBRIUM CONDITIONS

As long as there are no changes in the reaction conditions, a chemical reaction will maintain equilibrium. However, if a reaction that is at equilibrium is disturbed externally, this may affect its equilibrium. Le Chatelier's principle states that if chemical equilibrium is disturbed, the reaction will respond and attempt to eliminate or minimize the external disturbance.

Changing Concentration

Adding or removing reactants and products that are at a constant temperature in a chemical equilibrium disturbs the equilibrium, but it does not change the numerical value of K. Adding or removing pure solids or liquids has no effect on equilibrium because their concentrations do not really change during a chemical reaction.

An additional amount of reactants increases the value of the denominator in K. The reaction immediately shifts to the right, making more products to keep the value of K constant and establishing a new equilibrium.

At the new equilibrium, the concentrations of the reactants and products are different to the original equilibrium. However, their ratio in K remains constant due to the law of mass action.

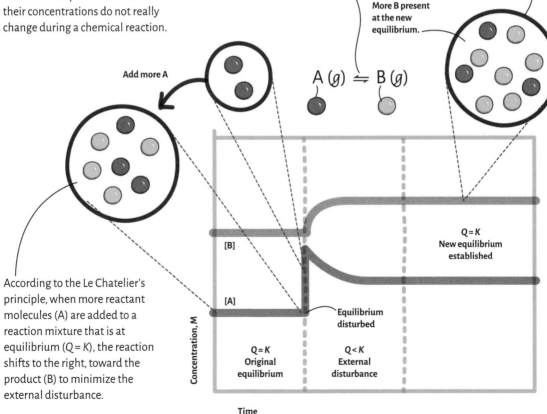

Add more A

More B present at the new equilibrium.

$A\,(g) \rightleftharpoons B\,(g)$

$Q = K$
New equilibrium established

According to the Le Chatelier's principle, when more reactant molecules (A) are added to a reaction mixture that is at equilibrium ($Q = K$), the reaction shifts to the right, toward the product (B) to minimize the external disturbance.

[B]

[A]

Equilibrium disturbed

Concentration, M

$Q = K$
Original equilibrium

$Q < K$
External disturbance

Time

Changing Pressure or Volume

As liquids and solids are incompressible, changes in pressure and volume will only effect gaseous equilibrium reactions. Pressure and volume are inversely related, so when one increases, the other one decreases.

The reaction between hydrogen (H_2) and nitrogen (N_2) gas produces gaseous ammonia (NH_3). An increase in pressure (and a decrease in volume) at a constant temperature will disturb equilibrium.

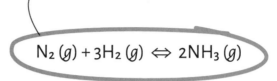

$$N_2\ (g) + 3H_2\ (g) \Leftrightarrow 2NH_3\ (g)$$

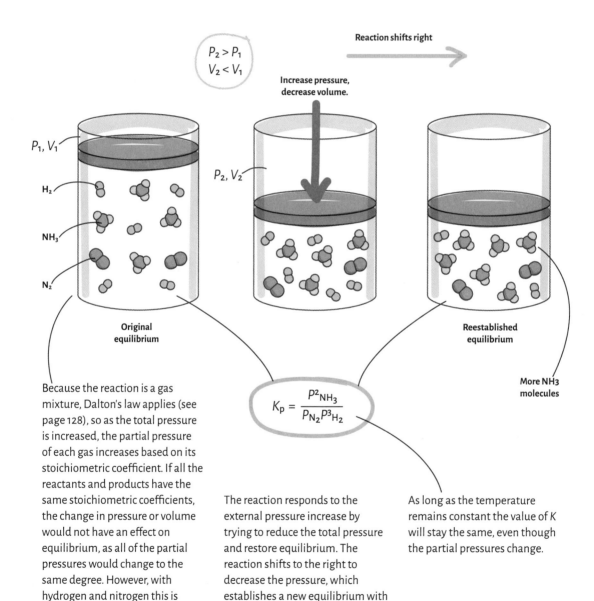

Reaction shifts right

$P_2 > P_1$
$V_2 < V_1$

Increase pressure, decrease volume.

P_1, V_1

H_2

NH_3

N_2

P_2, V_2

Original equilibrium

Reestablished equilibrium

More NH_3 molecules

$$K_p = \frac{P^2_{NH_3}}{P_{N_2}P^3_{H_2}}$$

Because the reaction is a gas mixture, Dalton's law applies (see page 128), so as the total pressure is increased, the partial pressure of each gas increases based on its stoichiometric coefficient. If all the reactants and products have the same stoichiometric coefficients, the change in pressure or volume would not have an effect on equilibrium, as all of the partial pressures would change to the same degree. However, with hydrogen and nitrogen this is not the case, so equilibrium is disturbed.

The reaction responds to the external pressure increase by trying to reduce the total pressure and restore equilibrium. The reaction shifts to the right to decrease the pressure, which establishes a new equilibrium with a higher ammonia concentration.

As long as the temperature remains constant the value of K will stay the same, even though the partial pressures change.

Temperature Change

Often, the temperature of the reactants in a chemical reaction needs to be increased for the reaction to occur. These are called **endothermic** (heat in) reactions, in which heat energy can be seen as a reactant that is required to start the reaction. **Exothermic** (heat out) reactions release heat energy as a product.

The decomposition reaction of N_2O_4 is an endothermic reaction. Increasing the temperature of this reaction at equilibrium means adding heat to the reactant side as an external disturbance. According to Le Chatelier's principle, the reaction shifts toward the product side to dissipate the added heat, which produces more NO_2.

A temperature increase moves a chemical equilibrium in the endothermic direction, while an exothermic reaction moves in the reverse direction, toward the reactants. Therefore, if heat is removed from the decomposition reaction of N_2O_4 at equilibrium, the reaction will proceed in the reverse direction, producing more N_2O_4 in order to attain a new equilibrium.

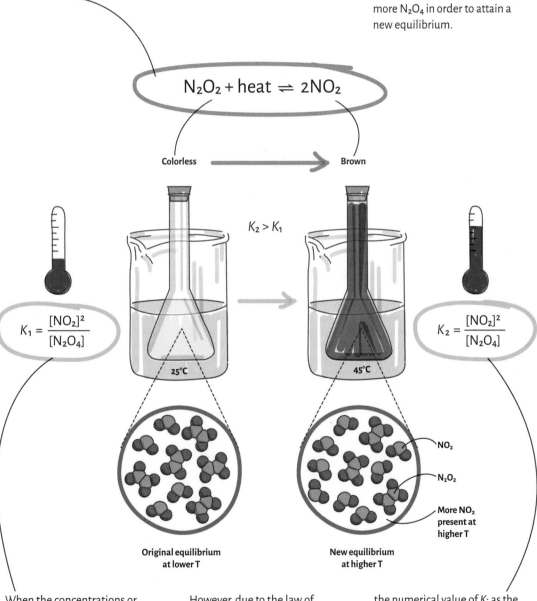

$$N_2O_2 + heat \rightleftharpoons 2NO_2$$

Colorless ⟶ Brown

$K_2 > K_1$

$$K_1 = \frac{[NO_2]^2}{[N_2O_4]}$$

25°C

45°C

$$K_2 = \frac{[NO_2]^2}{[N_2O_4]}$$

NO_2

N_2O_2

More NO_2 present at higher T

Original equilibrium at lower T

New equilibrium at higher T

When the concentrations or pressure (volume) are changed in a chemical equilibrium, a shift occurs in the reaction as it establishes a new equilibrium.

However, due to the law of mass action the numerical value of K does not change. A change in temperature will disturb the equilibrium and change

the numerical value of K; as the temperature increases, so does the value of K.

EQUILIBRIUM CALCULATIONS

As long as equilibrium exists, the concentrations of the reaction species remain unchanged, which allows various mathematical calculations to be carried out. If at least one of the equilibrium concentrations is known, the value of K can be calculated at the equilibrium temperature, and if K is known, the equilibrium concentrations of all species can be determined.

ICE Table

To make calculations simple, a grid is constructed that shows the initial concentrations (I), changes in concentrations (C), and equilibrium concentrations (E). The E row in an ICE table provides the information needed for equilibrium calculations involving K; the C row shows the drop in concentrations of reactants (–) and the increase in the concentrations of products (+); and x represents the unknown amounts of concentration changes of each species, based on their stoichiometric coefficients.

$$N_2\,(g) + 3H_2\,(g) \rightleftharpoons 2NH_3\,(g)$$

I	2	1	0
C			
E			

The initial concentrations are written in the I row of the table.

$$N_2\,(g) + 3H_2\,(g) \rightleftharpoons 2NH_3\,(g)$$

I	Initial concentrations (M)
C	Changes in concentration
E	Equilibrium concentrations (M)

The initial concentrations are written in the I row of the table. If the equilibrium concentration of any of the reaction species is known, it is written in the E row.

$$N_2\,(g) + 3H_2\,(g) \rightleftharpoons 2NH_3\,(g)$$

I	2	1	0
C			
E			0.5

FINDING EQUILIBRIUM CONCENTRATIONS
K is known, find equilibrium concentrations.

FINDING THE VALUE OF K
Equilibrium concentrations are known, find K.

The concentrations from the C row—given in terms of x—are plugged into the K expression. As the value of K is known, x can be determined, providing the numerical values for the equilibrium concentrations.

$$N_2\ (g) + 3H_2\ (g) \rightleftharpoons 2NH_3\ (g)$$

I	2	1	0
C	-x	-3x	+2x
E			

The changes in concentrations of all reactant species are entered in the C row in terms of x and the stoichiometric coefficients.

The E row is the sum of the I and C rows, and indicates the equilibrium conditions.

$$N_2\ (g) + 3H_2\ (g) \rightleftharpoons 2NH_3\ (g)$$

I	2	1	0
C	-x	-3x	+2x
E	2-x	1-3x	2x

$$K = \frac{(2x)^2}{(2-x)(1-3x)^3}$$

The rest of the grid can be completed using the stoichiometry of the reaction.

$$K = \frac{(0.5)^2}{(1.75)(0.25)^3}$$

$$N_2\ (g) + 3H_2\ (g) \rightleftharpoons 2NH_3\ (g)$$

I	2	1	0
C	-0.25	-0.75	0.25
E	1.75	0.25	0.5

The concentration values from the E row are entered into the K expression, enabling us to determine its numerical value at the reaction temperature.

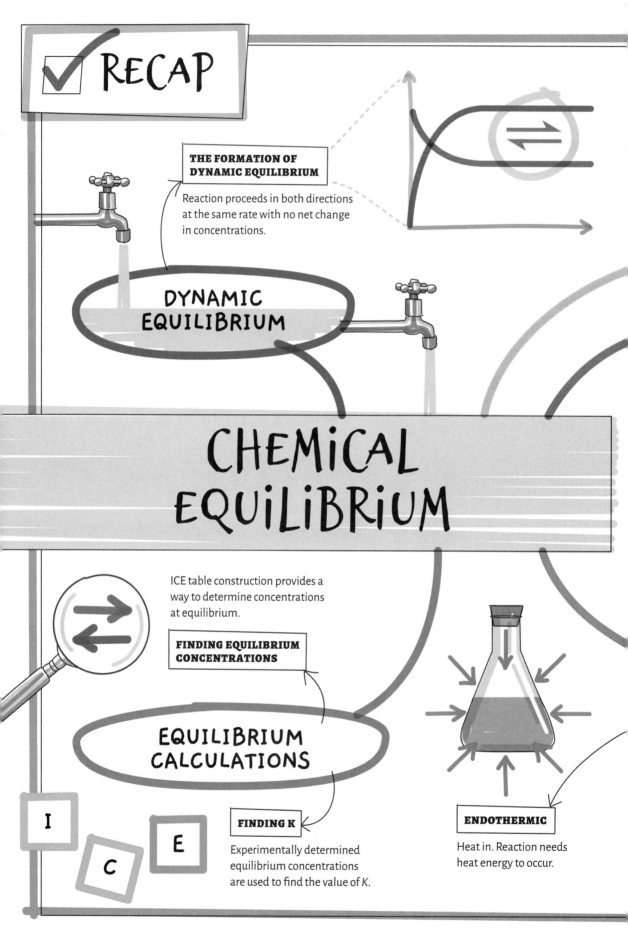

THE FORMATION OF DYNAMIC EQUILIBRIUM

Reaction proceeds in both directions at the same rate with no net change in concentrations.

DYNAMIC EQUILIBRIUM

CHEMICAL EQUILIBRIUM

ICE table construction provides a way to determine concentrations at equilibrium.

FINDING EQUILIBRIUM CONCENTRATIONS

EQUILIBRIUM CALCULATIONS

I
C
E

FINDING K

Experimentally determined equilibrium concentrations are used to find the value of K.

ENDOTHERMIC

Heat in. Reaction needs heat energy to occur.

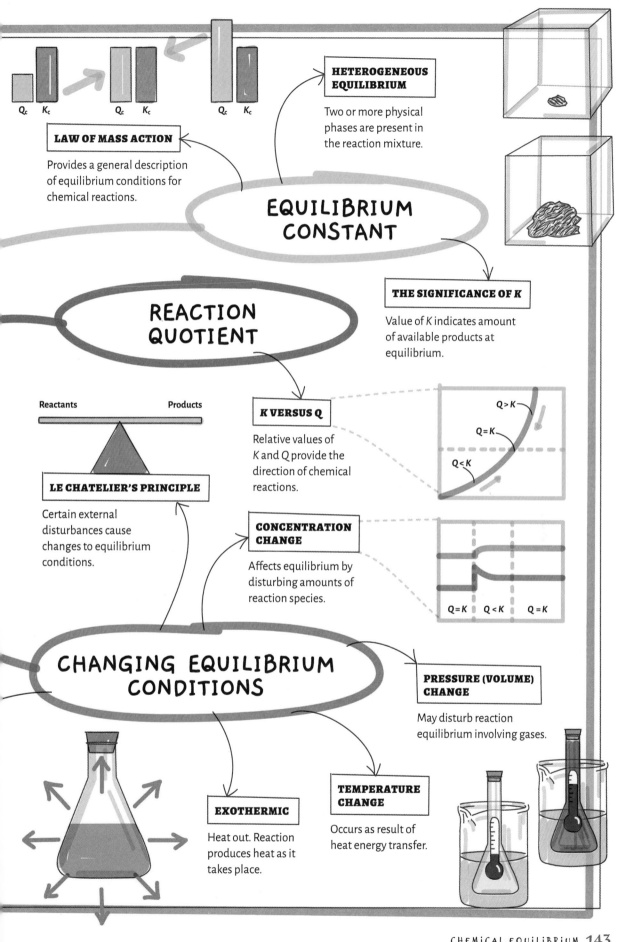

HETEROGENEOUS EQUILIBRIUM

Two or more physical phases are present in the reaction mixture.

Q_c K_c Q_c K_c Q_c K_c

LAW OF MASS ACTION

Provides a general description of equilibrium conditions for chemical reactions.

EQUILIBRIUM CONSTANT

REACTION QUOTIENT

THE SIGNIFICANCE OF K

Value of K indicates amount of available products at equilibrium.

Reactants   Products

K VERSUS Q

Relative values of K and Q provide the direction of chemical reactions.

$Q > K$

$Q = K$

$Q < K$

LE CHATELIER'S PRINCIPLE

Certain external disturbances cause changes to equilibrium conditions.

CONCENTRATION CHANGE

Affects equilibrium by disturbing amounts of reaction species.

$Q = K$ $Q < K$ $Q = K$

CHANGING EQUILIBRIUM CONDITIONS

PRESSURE (VOLUME) CHANGE

May disturb reaction equilibrium involving gases.

EXOTHERMIC

Heat out. Reaction produces heat as it takes place.

TEMPERATURE CHANGE

Occurs as result of heat energy transfer.

CHAPTER 12

ACIDS AND BASES

The terms *acid* and *base* are used to describe the characteristic properties of a major group of substances in chemistry. The name "acid" stems from the Latin, *acere*, meaning "sour," while "base" derives from the Arabic, *alqali*, which means "alkaline." Acids and bases find many applications in daily life: they play an important role in digesting the food we eat, in the effectiveness of many medicines we take, in the unique taste and smell of the various foods and beverages we enjoy, and in the function of household cleaning products.

DEFINING ACIDS AND BASES

The acidic or basic character of a substance is most frequently observed when it is mixed with water, where it can change the balance of hydrogen (H^+) ions and hydroxide (OH^-) ions. Substances that increase the hydrogen ion concentration in their aqueous solutions are **acids**, while substances that increase the hydroxide ion concentration behave as **bases**. **Amphoteric** substances can do both and act as an acid or a base depending upon their surroundings.

Autoionization of Water

Water ionizes to form **hydronium** (H_3O^+) and hydroxide ions. This happens when one water molecule collides with another and a hydrogen ion. The ionization reaction does not happen to an appreciable degree at room temperature and water molecules remain largely intact. Equilibrium exists between the reactants and products of the autoionization reaction of water.

At room temperature, the concentration of hydronium and hydroxide ions in water is very low (equal to 1.0×10^{-7} M). The equilibrium constant (K_w) therefore has a value of 1.0×10^{-14}. Since K_w is so small, the equilibrium for the autoionization of water lies far to the left, where water molecules are present.

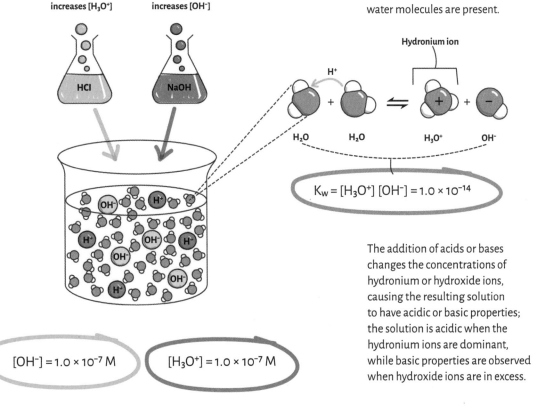

Adding acid increases [H_3O^+]

Adding base increases [OH^-]

HCl

NaOH

Hydronium ion

H^+

H_2O H_2O H_3O^+ OH^-

$$K_w = [H_3O^+][OH^-] = 1.0 \times 10^{-14}$$

$[OH^-] = 1.0 \times 10^{-7}$ M

$[H_3O^+] = 1.0 \times 10^{-7}$ M

The addition of acids or bases changes the concentrations of hydronium or hydroxide ions, causing the resulting solution to have acidic or basic properties; the solution is acidic when the hydronium ions are dominant, while basic properties are observed when hydroxide ions are in excess.

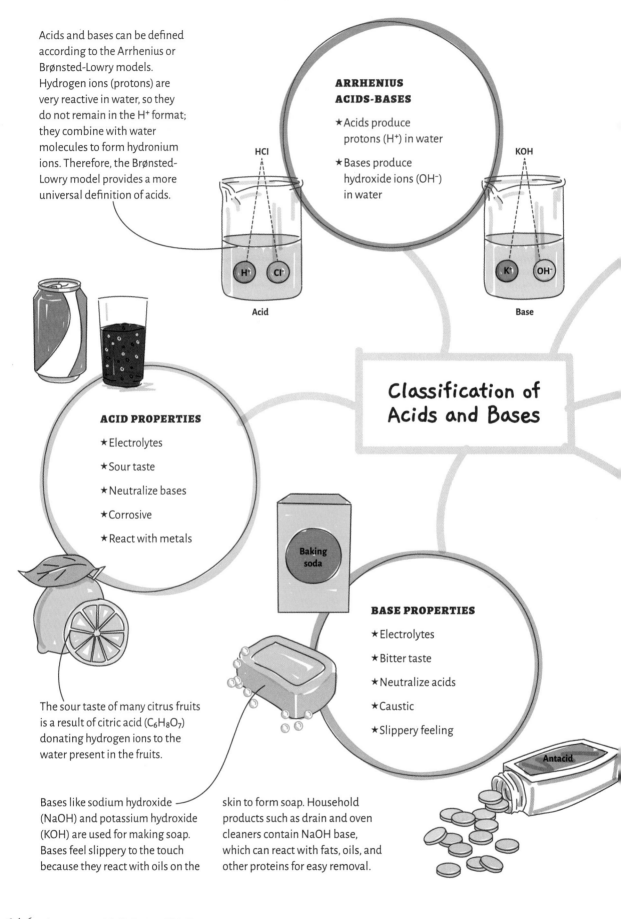

Acids and bases can be defined according to the Arrhenius or Brønsted-Lowry models. Hydrogen ions (protons) are very reactive in water, so they do not remain in the H⁺ format; they combine with water molecules to form hydronium ions. Therefore, the Brønsted-Lowry model provides a more universal definition of acids.

ARRHENIUS ACIDS-BASES

★ Acids produce protons (H^+) in water

★ Bases produce hydroxide ions (OH^-) in water

HCl

H^+ Cl^-

Acid

KOH

K^+ OH^-

Base

Classification of Acids and Bases

ACID PROPERTIES

★ Electrolytes

★ Sour taste

★ Neutralize bases

★ Corrosive

★ React with metals

Baking soda

BASE PROPERTIES

★ Electrolytes

★ Bitter taste

★ Neutralize acids

★ Caustic

★ Slippery feeling

Antacid

The sour taste of many citrus fruits is a result of citric acid ($C_6H_8O_7$) donating hydrogen ions to the water present in the fruits.

Bases like sodium hydroxide (NaOH) and potassium hydroxide (KOH) are used for making soap. Bases feel slippery to the touch because they react with oils on the skin to form soap. Household products such as drain and oven cleaners contain NaOH base, which can react with fats, oils, and other proteins for easy removal.

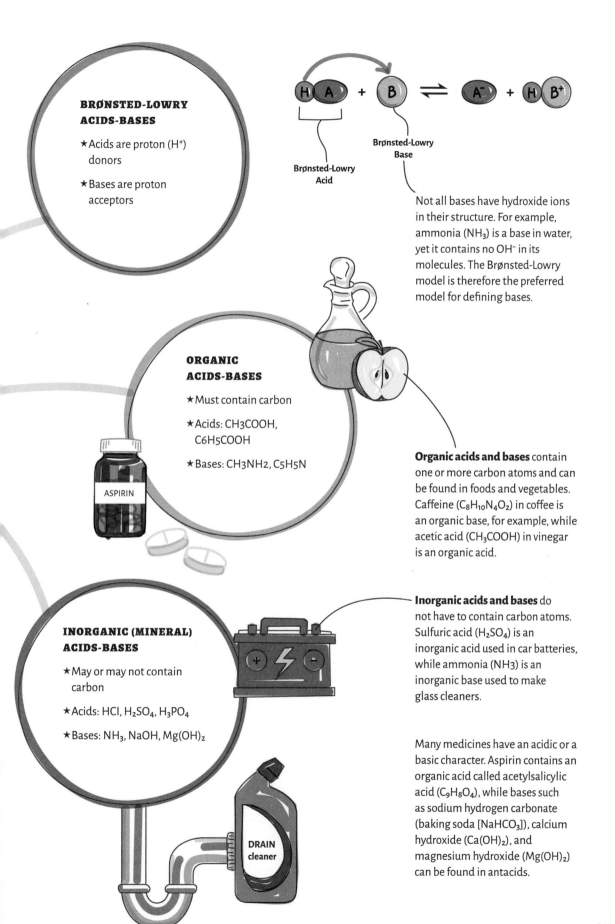

BRØNSTED-LOWRY ACIDS-BASES

★ Acids are proton (H^+) donors

★ Bases are proton acceptors

Brønsted-Lowry Acid

Brønsted-Lowry Base

Not all bases have hydroxide ions in their structure. For example, ammonia (NH_3) is a base in water, yet it contains no OH^- in its molecules. The Brønsted-Lowry model is therefore the preferred model for defining bases.

ORGANIC ACIDS-BASES

★ Must contain carbon

★ Acids: CH3COOH, C6H5COOH

★ Bases: CH3NH2, C5H5N

ASPIRIN

Organic acids and bases contain one or more carbon atoms and can be found in foods and vegetables. Caffeine ($C_8H_{10}N_4O_2$) in coffee is an organic base, for example, while acetic acid (CH_3COOH) in vinegar is an organic acid.

Inorganic acids and bases do not have to contain carbon atoms. Sulfuric acid (H_2SO_4) is an inorganic acid used in car batteries, while ammonia (NH_3) is an inorganic base used to make glass cleaners.

INORGANIC (MINERAL) ACIDS-BASES

★ May or may not contain carbon

★ Acids: HCl, H_2SO_4, H_3PO_4

★ Bases: NH_3, NaOH, $Mg(OH)_2$

DRAIN cleaner

Many medicines have an acidic or a basic character. Aspirin contains an organic acid called acetylsalicylic acid ($C_9H_8O_4$), while bases such as sodium hydrogen carbonate (baking soda [$NaHCO_3$]), calcium hydroxide ($Ca(OH)_2$), and magnesium hydroxide ($Mg(OH)_2$) can be found in antacids.

THE pH SCALE

The level of acidity or alkalinity of an acid-base solution is quantitatively reported by the pH (power of hydrogen) or pOH (power of hydroxide). These are both logarithmic scales, referring to the concentrations of hydronium ions and hydroxide ions, respectively.

pH and pOH

pH is defined as the negative log of hydronium ion concentration in molarity. pH values range from 0 to 14 in aqueous solutions, making it a convenient way to numerically describe how acidic a solution is.

pOH is the negative log of hydroxide ion concentration in molarity. It provides a quantitative way of reporting the level of alkalinity in an aqueous solution; like pH it has a range of 0 to 14.

Neutral	Acidic	Basic
pH = 7	pH < 7	pH > 7
pOH = 7	pOH > 7	pOH < 7

Aqueous solutions are classified as acidic, basic, or neutral, based on their pH (or pOH) value. Pure water is neutral because it has a pH of 7 and a pOH of 7. When an acidic substance is added to water, the hydronium ion concentration increases, causing the pH to drop below 7. A basic substance, on the other hand, has an increased hydroxide ion concentration, which raises the pH above 7.

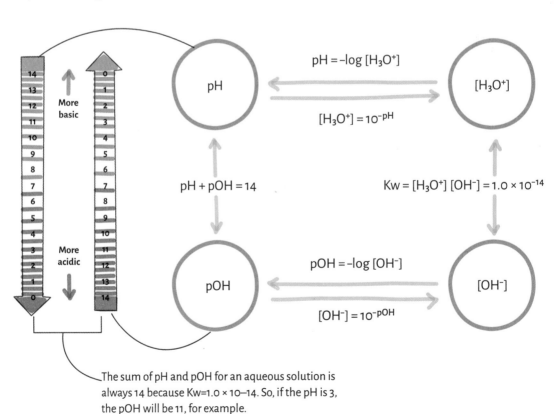

$$pH = -\log [H_3O^+]$$

$$[H_3O^+] = 10^{-pH}$$

$$pH + pOH = 14$$

$$Kw = [H_3O^+][OH^-] = 1.0 \times 10^{-14}$$

$$pOH = -\log [OH^-]$$

$$[OH^-] = 10^{-pOH}$$

The sum of pH and pOH for an aqueous solution is always 14 because Kw=1.0 × 10–14. So, if the pH is 3, the pOH will be 11, for example.

Acids and Bases in Daily Life

Many of the foods, drinks, medicines, and household products we use every day have acidic or alkaline properties. Without their characteristic properties as proton donors or acceptors these acids and bases could not provide the functions that we depend on. Antacids, for example, work well in alleviating acid reflux because of their ability to remove excess protons from the gastric juice.

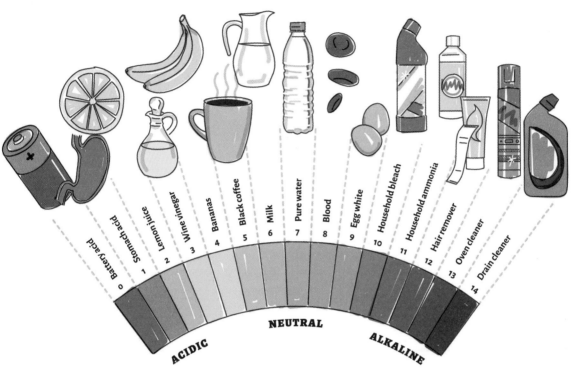

Various pH levels exist in different parts of the human body. For example, the gastric juice in our stomach has a low pH and is highly corrosive, but without it we would not be able to digest food. Specific pH levels must be maintained in different organs, which requires a well-balanced diet; the frequent consumption of foods and drinks that are either too acidic or basic can be harmful to our health.

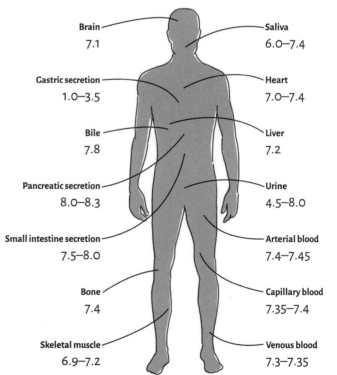

Brain
7.1

Saliva
6.0–7.4

Gastric secretion
1.0–3.5

Heart
7.0–7.4

Bile
7.8

Liver
7.2

Pancreatic secretion
8.0–8.3

Urine
4.5–8.0

Small intestine secretion
7.5–8.0

Arterial blood
7.4–7.45

Bone
7.4

Capillary blood
7.35–7.4

Skeletal muscle
6.9–7.2

Venous blood
7.3–7.35

Acid–Base Strength

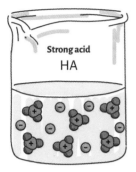

Strong acid
HA

$$HA\,(aq) + H_2O\,(l) \rightarrow H_3O^+\,(aq) + A^-\,(aq)$$

Some acids and bases completely ionize in water to form strong electrolyte solutions. These are called **strong acids and bases**. Hydrochloric acid (HCl) is a strong acid, which donates all of its protons to water to ionize completely. HCl is present in gastric juice and because of its strength it can break down food very easily.

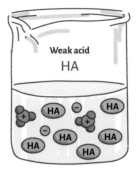

Weak acid
HA

$$K_a = \frac{[H_3O^+]\,[A^-]}{[HA]} \ll 1$$

$$HA\,(aq) + H_2O\,(l) \rightleftharpoons H_3O^+\,(aq) + A^-\,(aq)$$

Weak acids and bases largely stay intact and do not ionize to an appreciable amount in water, creating weak electrolyte solutions. They form an equilibrium with water, but this equilibrium lies far to the left with a small equilibrium constant. Acetic acid (CH_3COOH) in vinegar is a weak acid that does not free up a significant number of its protons. This means that it mainly stays in the CH_3COOH form when it is dissolved in water, which is why vinegar can be used as a salad dressing.

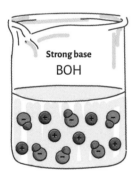

Strong base
BOH

$$BOH\,(aq) \rightarrow B^+\,(aq) + OH^-\,(aq)$$

Sodium hydroxide (NaOH) is a strong base because it completely ionizes in water to release all of its sodium and hydroxide ions. The high concentration of hydroxide ions makes NaOH highly caustic, and capable of breaking down biological matter. For this reason it is often used to open clogged drains.

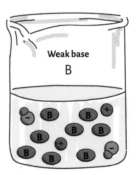

Weak base
B

$$K_b = \frac{[BH^+]\,[OH^-]}{[B]} \ll 1$$

$$B\,(aq) + H_2O\,(l) \rightleftharpoons BH^+\,(aq) + OH^-\,(aq)$$

Ammonia (NH_3) is a weak base that is incapable of accepting many protons from water, so only a limited number of hydroxide ions are created in solution. This is why ammonia can be used safely in household cleaning agents without the caustic effects of strong bases.

ACID-BASE INDICATORS

Some complex organic molecules, called **acid-base indicators** or **pH indicators**, are sensitive to changes in pH. These compounds are themselves weak acids and bases with limited ionization in water.. When they are in solution these indicators exhibit a distinct color when a proton attaches to them and another color when a proton detaches itself from the molecule, so they show a different color depending on pH. Such indicators are very useful when a quick check of pH levels is desired, so they have many applications. Nature is also full of examples of indicator molecules.

pH indicators can be made artificially or they can occur naturally. Anthocyanin is a molecule that appears in purple cabbage. It is capable of changing color in response to varying pH values.

Anthocyanins also exist in hydrangeas, giving the flowers a blue color if the plant is grown in acidic soil and a pink-purple color if it is planted in basic soil.

Acidic soil

Alkaline soil

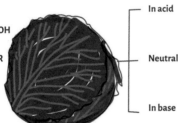

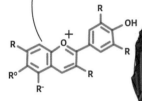

In acid

Neutral

In base

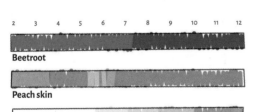

2 3 4 5 6 7 8 9 10 11 12

Beetroot

Peach skin

Tomato

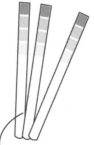

pH paper is a convenient tool for performing a quick check for acidity levels in various environments, such as swimming pools. It is made by coating paper with a universal indicator.

There are many examples of edible pH indicators in nature. In each case a specific, naturally occurring complex molecule provides various colors that indicate changes in pH.

Synthetic and natural molecules can be used to manufacture commercially available pH indicators. The most commonly used indicators are **universal acid-base indicators**, which are a blend of several molecules that provide a color spectrum for the entire pH range.

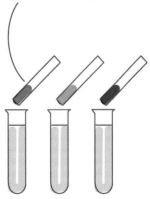

NEUTRALIZATION REACTIONS

Acids and bases have opposing properties, especially in terms of their affinity to hydrogen ions, which means they react with each other to form water and some form of ionic compound or salt. The hydrogen ions in acids and the hydroxide ions in bases react to form water, which is why the acid-base interaction is called a neutralization reaction. There are many examples in daily life that utilize **neutralization reactions**.

Neutralization

Acids neutralize the alkaline properties of bases. The ionic compound that forms will depend on the type of acid and base used in the neutralization reaction. When sodium hydroxide (NaOH) and hydrochloric acid (HCl) react, for example, they form sodium chloride (NaCl) and water (H_2O).

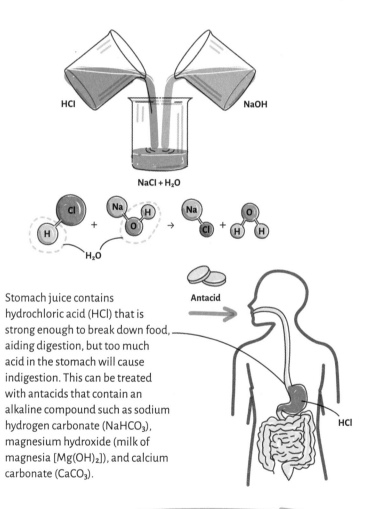

Ant bites and bee stings are acidic because of the excretion of formic acid (HCOOH). Sodium hydrogen carbonate ($NaHCO_3$), which is a base, can be used to neutralize the effect. However, a wasp sting contains compounds with basic properties, so a vinegar containing acetic acid is usually used as a treatment.

Stomach juice contains hydrochloric acid (HCl) that is strong enough to break down food, aiding digestion, but too much acid in the stomach will cause indigestion. This can be treated with antacids that contain an alkaline compound such as sodium hydrogen carbonate ($NaHCO_3$), magnesium hydroxide (milk of magnesia [$Mg(OH)_2$]), and calcium carbonate ($CaCO_3$).

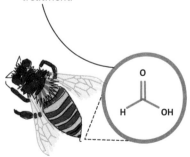

$$2HCl \ (aq) + Mg(OH)_2 \ (aq) \rightarrow 2H_2O \ (l) + MgCl_2 \ (aq)$$

$$HCOOH \ (aq) + NaHCO_3 \ (aq) \rightarrow NaCOOH \ (aq) + H_2O \ (l) + CO_2 \ (g)$$

Acid Rain

Nonmetal oxides, such as sulfur dioxide (SO_2), nitric oxide (NO), nitrogen dioxide (NO_2), and carbon dioxide (CO_2), form acidic solutions when mixed with water. They are called **acidic anhydrides** and are found in the polluted air emitted by industrial plants, automobiles, and volcanoes.

Carbon dioxide (CO_2) combines with rainwater to form a weak solution of carbonic acid (H_2CO_3), which is why normal rainwater is acidic, with a pH of approximately 5.6.

Pure rain drops
pH = 7

H_2O

Nitric / nitrous acid
$N_2 + O_2 \rightarrow 2NO$
$2NO + O2 \rightarrow 2NO_2$

$S + O_2 \rightarrow SO_2$

Sulfurous acid
pH < 6

Carbonic acid
pH < 7
CO_2

H_2CO_3

H_2SO_3

$SO_2 + O_2 \rightarrow SO_3$

HNO_3
HNO_2

Sulfuric acid
pH < 5

Pollution
SO_2 NO_x

H_2SO_4

Gases like SO_2, NO, and NO_2 make it into the atmosphere through air pollution. When these gases dissolve in rainwater, strong acidic conditions are formed, causing the pH to drop significantly below 6. This effect is called acid rain. The pH of rainwater in the northeast of the United States can be as low as 4.2.

Pollution
NO_x

Acidic rain water

Acid rain causes significant economic, health, and environmental damage, acidifying soil and water, which threatens aquatic life and vegetation. Each year, millions of tons of lime (CaO) are added to soil, lakes, and rivers around the world to counteract acidification due to acid rain. The lime reacts with water to form calcium hydroxide ($Ca(OH)_2$), which is a base that can neutralize acidic compounds in soil and water.

Acid-Base Titration

Neutralization reactions are often used to perform quantitative chemical analysis to determine the concentration of an acid or a base in a sample. This process is called acid-base titration.

An acid sample, for example, is placed into a flask, along with a few drops of pH indicator. A suitable base solution with a known concentration is then added slowly to the acid sample using a buret. The pH change in the flask is often monitored with a digital pH-meter.

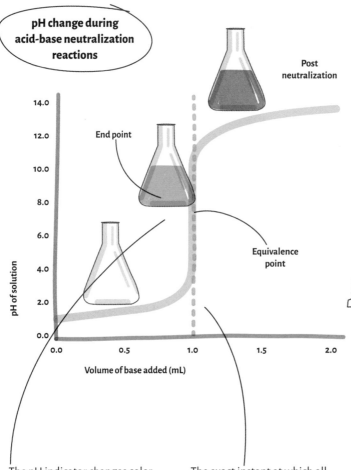

Acid-base titration

Buret containing base solution.

pH change during acid-base neutralization reactions

Post neutralization

End point

Equivalence point

Acid sample

pH of solution

Volume of base added (mL)

pH

The pH indicator changes color when all of the acid has been neutralized by the added base; this is called the **end point**. For this color change to occur the reaction needs to pass the exact neutralization point to create slightly basic conditions.

The exact instant at which all the acid is neutralized by the base is called the **equivalence point**. At the equivalence point all of the hydrogen ions from the acid have been neutralized by the hydroxide ions from the base, forming water.

The volume of added base at the equivalence point is used to stoichiometrically determine the concentration of the acid sample.

BUFFERS

Aqueous solutions capable of resisting changes in pH when acids or bases are added to them are called **buffer** solutions. Buffers consist of a weak acid and its salt with a strong base, or a weak base and its salt with a strong acid. Many biological systems are sensitive to pH changes in their environment, so buffers play a crucial role in maintaining a healthy pH balance for normal biological activity.

Buffering Effect

Carbonic acid (H_2CO_3) is a weak acid that can form a buffer solution when mixed with its salt (sodium hydrogen carbonate [$NaHCO_3$], for example). The bicarbonate ion (HCO_3^-) is the base form of carbonic acid and is provided by the salt; this is called the **conjugate base**. When the H_2CO_3/HCO_3^- pair is present in an aqueous solution, the buffering effect occurs and the solution resists changes in pH.

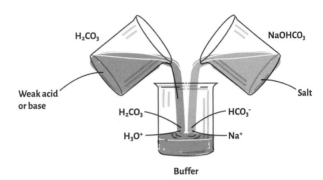

If a base is added to the solution containing the carbonic acid/bicarbonate buffer pair, H_2CO_3 immediately neutralizes it before the pH value increases. Similarly, the addition of an acid is neutralized by HCO_3^- to keep the pH constant.

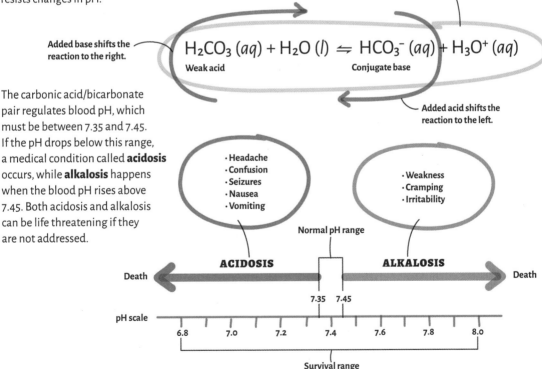

Added base shifts the reaction to the right.

$$H_2CO_3\ (aq) + H_2O\ (l) \rightleftharpoons HCO_3^-\ (aq) + H_3O^+\ (aq)$$

Weak acid   Conjugate base

Added acid shifts the reaction to the left.

The carbonic acid/bicarbonate pair regulates blood pH, which must be between 7.35 and 7.45. If the pH drops below this range, a medical condition called **acidosis** occurs, while **alkalosis** happens when the blood pH rises above 7.45. Both acidosis and alkalosis can be life threatening if they are not addressed.

- Headache
- Confusion
- Seizures
- Nausea
- Vomiting

- Weakness
- Cramping
- Irritability

Normal pH range

ACIDOSIS   **ALKALOSIS**

Death   Death

7.35   7.45

pH scale

6.8   7.0   7.2   7.4   7.6   7.8   8.0

Survival range

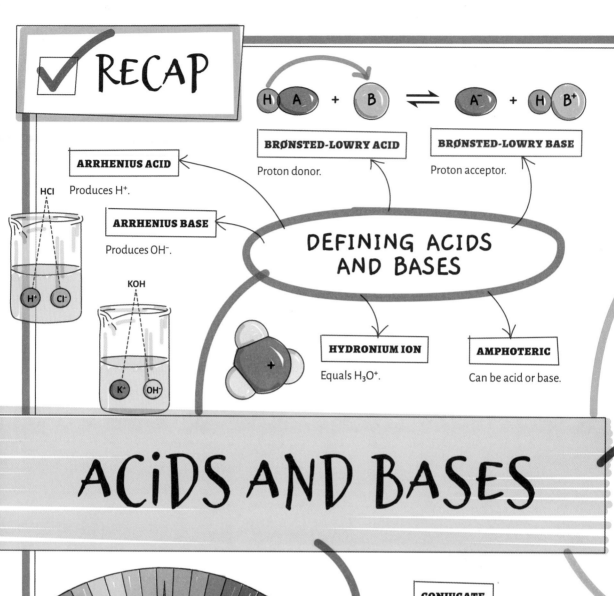

RECAP

BRØNSTED-LOWRY ACID

Proton donor.

BRØNSTED-LOWRY BASE

Proton acceptor.

ARRHENIUS ACID

Produces H^+.

ARRHENIUS BASE

Produces OH^-.

HCl

DEFINING ACIDS AND BASES

HYDRONIUM ION

Equals H_3O^+.

AMPHOTERIC

Can be acid or base.

KOH

ACIDS AND BASES

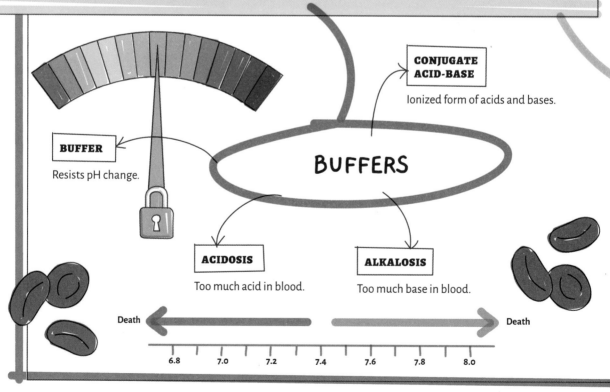

CONJUGATE ACID-BASE

Ionized form of acids and bases.

BUFFER

Resists pH change.

BUFFERS

ACIDOSIS

Too much acid in blood.

ALKALOSIS

Too much base in blood.

Death ← → Death

6.8 7.0 7.2 7.4 7.6 7.8 8.0

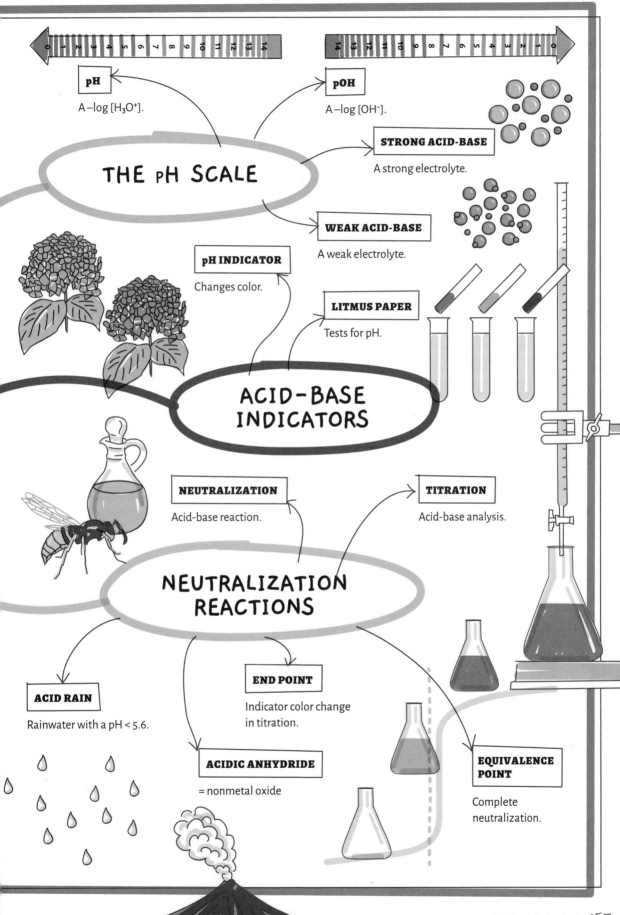

pH

A –log [H_3O^+].

pOH

A –log [OH^-].

THE pH SCALE

STRONG ACID-BASE

A strong electrolyte.

WEAK ACID-BASE

A weak electrolyte.

pH INDICATOR

Changes color.

LITMUS PAPER

Tests for pH.

ACID–BASE INDICATORS

NEUTRALIZATION

Acid-base reaction.

TITRATION

Acid-base analysis.

NEUTRALIZATION REACTIONS

ACID RAIN

Rainwater with a pH < 5.6.

END POINT

Indicator color change in titration.

ACIDIC ANHYDRIDE

= nonmetal oxide

EQUIVALENCE POINT

Complete neutralization.

CHAPTER 13

THERMODYNAMICS

The term thermodynamics is derived from the words *thermo*, meaning "heat," and *dynamics*, meaning "motion." As the name suggests, this important branch of science deals with heat and other forms of energy, and the relationships between them. It is essentially the science of the energy changes that occur within physical and chemical processes when energy is transferred from one place to another, or from one form to another. As such, the laws of thermodynamics are among the most fundamental in all of science, as they provide an explanation about what ultimately drives physical and chemical changes in matter.

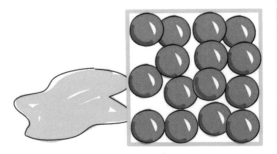

THERMODYNAMICS AND ENTHALPY

The **first law of thermodynamics** is fundamentally related to the natural law of energy conservation. It states that the total energy of the universe is constant, so energy cannot be destroyed or created during a physical or chemical change. It can, however, be converted from one form to another, or flow from one place to another.

Thermodynamic Standard State

Energy terms in thermodynamics are often expressed under standard conditions of 298.15 K temperature and 1 atm pressure when all substances involved are in their most stable form. Thermodynamic data for many chemical and physical processes are reported in the scientific literature at standard conditions.

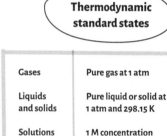

Thermodynamic standard states

Gases	Pure gas at 1 atm
Liquids and solids	Pure liquid or solid at 1 atm and 298.15 K
Solutions	1 M concentration

System and Surroundings

A thermodynamic **system** is defined as a small region that is under investigation. For example, the system could be a chemical reaction taking place in a test tube, or it could be a sheet of ice sitting on a table; the immediate environment around the system is the surroundings. The system and **surroundings** can exchange matter and energy and together make up the **universe**.

The **internal energy (E)** of a system is the sum of the kinetic and potential energies of the contents of that system. When a chemical or physical change occurs, energy is exchanged between the system and its surroundings. The change in system energy is represented by ΔE, which is the difference between the internal energies of the system after (E_{final}) and before ($E_{initial}$) the change took place.

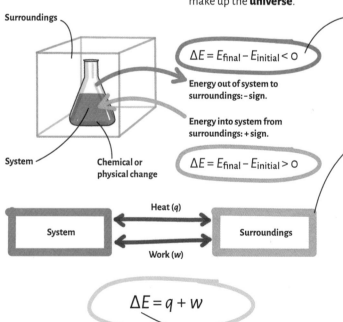

Surroundings

$$\Delta E = E_{final} - E_{initial} < 0$$

Energy out of system to surroundings: – sign.

Energy into system from surroundings: + sign.

System

Chemical or physical change

$$\Delta E = E_{final} - E_{initial} > 0$$

Heat (q)

| System | Surroundings |

Work (w)

$$\Delta E = q + w$$

The first law of thermodynamics states that the exchange of energy between the system and its surroundings can be in the form of **heat (q)** and **work (w)**.

ΔE is a **state function**, which means its value depends solely on the system's initial and final states, and not the nature of the path taken for the change to occur.

Enthalpy

Many chemical and physical processes take place under the constant atmospheric pressure conditions where we live. At constant pressure conditions the exchange of heat energy (q) due to a temperature difference between the system and its surroundings is called **enthalpy (H)**. This is also a state function.

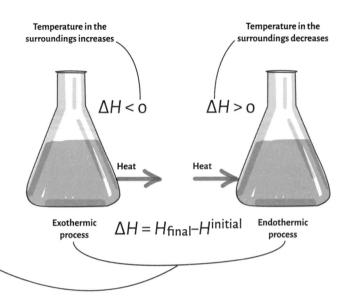

Temperature in the surroundings increases

Temperature in the surroundings decreases

$\Delta H < 0$

$\Delta H > 0$

Heat

Heat

Exothermic process

$\Delta H = H_{final} - H_{initial}$

Endothermic process

If the system loses heat to its surroundings ($\Delta H < 0$), the process that took place in the system is called **exothermic**. If the system gains heat from its surroundings ($\Delta H > 0$), then it is an **endothermic** process.

$$4Fe\ (s) + 3O_2\ (g) \rightarrow 2Fe_2O_3\ (s) + heat$$

There are many examples in daily life where we can see heat energy transferred between a system and its surrounding. Chemical hand warmers, for example, involve an exothermic reaction that releases heat from the system (hand warmer) to the surroundings (hands).

Hand warmers are made by sealing iron powder, water, and salt in a gas-permeable pouch. When the pouch is removed from its outer packaging, oxygen gas from the air permeates the pouch and reacts with the iron powder to make rust. This reaction is highly exothermic, which means it creates heat.

The pouch (system) is at a higher temperature than our cold hands (surroundings) when the reaction takes place. The heat that's released by the reaction flows out of the pouch and into our hands, creating the hand-warming effect. Over time the system loses heat, while the surroundings gain heat.

The amount of heat produced by the reaction can be determined from the standard enthalpy of formation (ΔH°_f) data provided in the scientific literature. The reaction in hand warmer pouches releases -1652 kJ of energy.

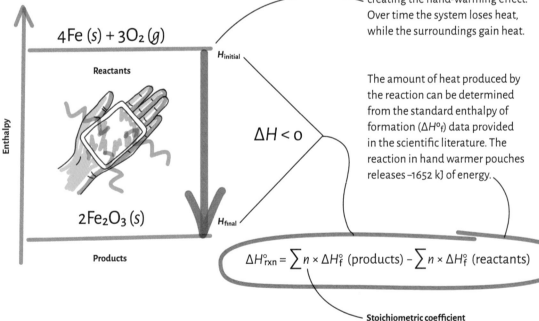

Enthalpy

$4Fe\ (s) + 3O_2\ (g)$

$H_{initial}$

Reactants

$\Delta H < 0$

$2Fe_2O_3\ (s)$

H_{final}

Products

$$\Delta H^{\circ}_{rxn} = \sum n \times \Delta H^{\circ}_f\ (products) - \sum n \times \Delta H^{\circ}_f\ (reactants)$$

Stoichiometric coefficient

Measuring Enthalpy

Calorimetry is a technique utilized by scientists to measure the heat exchange between a system and its surroundings. A thermometer is used to determine the temperature change in the surroundings, which provides a quantitative measure of how much heat flows during the process.

The enthalpy change for many aqueous reactions can be measured directly in a **constant pressure calorimeter.** Using this method, a chemical reaction takes place in an aqueous solution placed in a well-insulated container with a loose-fitting lid to ensure constant atmospheric pressure conditions. The temperature of the solution (which is acting as the surroundings) is measured before and after the reaction takes place. If the temperature of the solution increases, the reaction is exothermic, while a drop in temperature indicates an endothermic reaction.

The change in temperature (ΔT) is directly related to the enthalpy change for the reaction (ΔH).

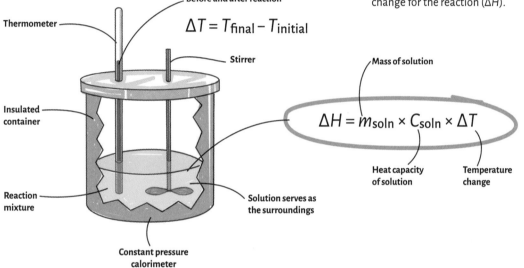

Temperature is measured before and after reaction

$$\Delta T = T_{final} - T_{initial}$$

Thermometer

Stirrer

Mass of solution

Insulated container

$$\Delta H = m_{soln} \times C_{soln} \times \Delta T$$

Heat capacity of solution

Temperature change

Reaction mixture

Solution serves as the surroundings

Constant pressure calorimeter

Enthalpy for Phase Transformations

The enthalpy change during phase transformations can easily be measured using calorimetry. Fusion (melting), vaporization, and sublimation are all endothermic processes that require heat to take place, so $\Delta H > 0$. Conversely, freezing, condensation, and deposition are all exothermic processes where $\Delta H < 0$.

The ambient temperature increases when it starts snowing, because the transformation of gaseous water into solid snow is an exothermic process releasing heat into the surroundings.

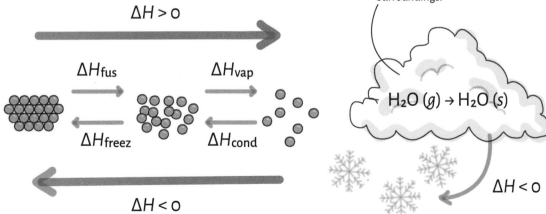

$$\Delta H > 0$$

$$\Delta H_{fus}$$

$$\Delta H_{vap}$$

$$\Delta H_{freez}$$

$$\Delta H_{cond}$$

$$\Delta H < 0$$

$$H_2O\ (g) \rightarrow H_2O\ (s)$$

$$\Delta H < 0$$

THERMODYNAMICS AND ENTROPY

The **second law of thermodynamics** is the most fundamental law of nature. It has profound implications that deal with entropy (S), which provides information about the disorder of a system. According to the second law, the entropy of the universe can never decrease when a spontaneous process takes place, and energy tends to spread out instead of being concentrated in one area.

Spontaneous and Nonspontaneous Processes

Once initiated, a spontaneous process is one that can proceed by itself without continuous outside interference. Any process that is **spontaneous** in one direction is **nonspontaneous** in the reverse direction.

Ice melts spontaneously, as long as the temperature of its surroundings is greater than the melting temperature of the ice. The direction of the process is from ice toward liquid water; the reverse process is impossible, as long as the temperature is above 0°C. Molecules in liquid water are more disorderly compared to their state in ice, resulting in the dispersal of energy.

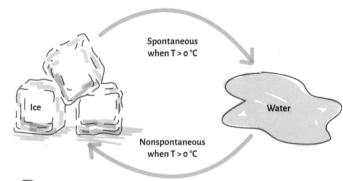

Spontaneous when T > 0°C

Ice

Water

Nonspontaneous when T > 0°C

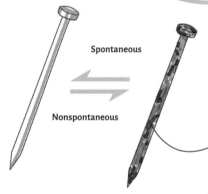

Spontaneous

Nonspontaneous

Natural processes are spontaneous. A nail will rust spontaneously because the entropy of the universe increases during this chemical change, and it is impossible to expect it to proceed in the reverse direction on its own.

Entropy

Entropy (S) can be described as a measure of the degree of randomness or freedom of the particles—such as atoms, ions, and molecules—in a system. Based on this definition, gases have higher entropy than liquids and solids because gas particles have much higher freedom, randomness, and disorder.

All phase changes from solid to liquid to gas involve an increase in entropy, while the change in entropy is negative in the reverse direction. The direction of increasing entropy allows energy to spread out.

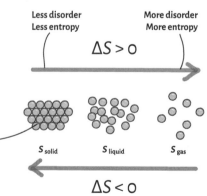

Less disorder
Less entropy

More disorder
More entropy

$\Delta S > 0$

S_{solid} S_{liquid} S_{gas}

$\Delta S < 0$

Entropy and Spontaneity

The exchange of enthalpy between the system and its surroundings inevitably affects the entropy. In an exothermic process, heat flows from the system into the surroundings, increasing the molecular disorder in the surroundings. This causes the entropy of the surroundings to increase. The opposite happens for an endothermic process.

A hot cup of coffee (system) is expected to lose heat to cooler surroundings. The heat gained by the surroundings is identical to the amount of heat lost by the cup. This allows us to determine the entropy change in the surroundings.

The coffee cup loses energy, and its contents get cooler over time. The decrease in temperature reduces the molecular disorder in the system, causing its entropy to decrease.

Because of the tendency for energy to spread out, hot coffee will lose heat to its cooler surroundings. This is the natural and spontaneous direction for the process, and as long as the conditions remain unchanged the reverse is impossible. The second law of thermodynamics quantitatively states that the entropy of the universe must increase, $\Delta S_{univ} > 0$, for a spontaneous process.

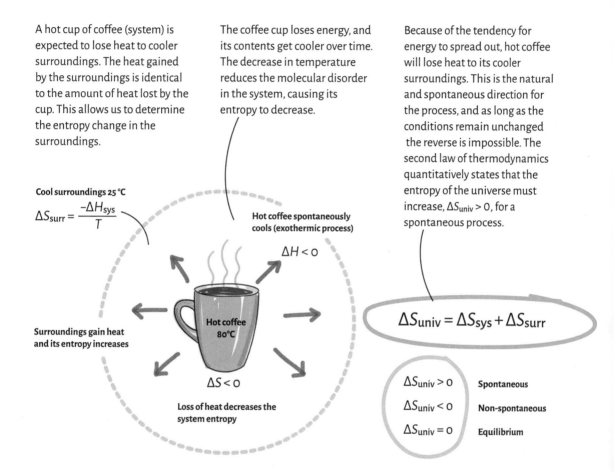

Cool surroundings 25 °C

$$\Delta S_{surr} = \frac{-\Delta H_{sys}}{T}$$

Hot coffee spontaneously cools (exothermic process)

$\Delta H < 0$

Hot coffee 80°C

Surroundings gain heat and its entropy increases

$\Delta S < 0$

Loss of heat decreases the system entropy

$$\Delta S_{univ} = \Delta S_{sys} + \Delta S_{surr}$$

$\Delta S_{univ} > 0$	**Spontaneous**
$\Delta S_{univ} < 0$	**Non-spontaneous**
$\Delta S_{univ} = 0$	**Equilibrium**

Enthalpy and entropy changes of a system alone can't quantitatively predict if the process occurring in the system is spontaneous or not. It is what happens to the entropy of the universe (system + surroundings, or ΔS_{univ}) that provides the criterion for spontaneity. The value of ΔS_{univ} is positive for spontaneous processes, is negative for nonspontaneous processes, and has a value of 0 for equilibrium processes.

Although the entropy of the coffee cup system is decreasing, the increase in the entropy of the surroundings is large enough to keep $\Delta S_{univ} > 0$. This ensures the cooling process remains spontaneous until thermal equilibrium is reached.

When the system involves a chemical reaction, the entropy change is easily determined from the standard entropy values ($S°$) of the reactant and product species available in the scientific literature.

$$\Delta S°_{rxn} = \sum n \times S° \text{ (products)} - \sum n \times S° \text{ (reactants)}$$

GiBBS FREE ENERGY AND SPONTANEiTY

The entropy change of the universe predicts whether a chemical reaction will occur spontaneously or not. However, this first requires knowledge about the entropy of the surroundings, which is not always easy to determine. **Gibbs free energy (G)** is a conveniently defined energy term that allows us to focus on the system properties alone when predicting the spontaneous nature of a chemical or physical change.

Gibbs Free Energy

Gibbs free energy (also called **chemical potential**) is a quantitative way of assessing the direction of change and the ability of a chemical or physical process to bring about that change. It is a system property defined in terms of system enthalpy and entropy alone, without the need for information about changes taking place in the surroundings.

The change in Gibbs energy (ΔG) provides a new set of criteria for the quantitative prediction of spontaneous change. When ΔG is negative, the process takes place spontaneously in the forward direction under the given set of temperature and pressure conditions. As long as the temperature and pressure remain constant, ΔG remains negative and the process continues to be spontaneous.

When $\Delta G = 0$, the process is said to be at a dynamic equilibrium. This means the forward and reverse processes are taking place spontaneously at the same rate, resulting in no overall change.

Processes with a negative ΔG are called **exergonic**, while **endergonic** processes result in a positive change in ΔG. A chemical or physical change is spontaneous in the exergonic direction and nonspontaneous in the endergonic direction.

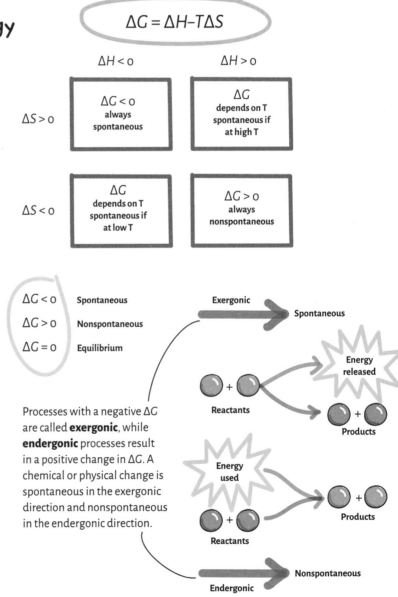

$$\Delta G = \Delta H - T\Delta S$$

	$\Delta H < 0$	$\Delta H > 0$
$\Delta S > 0$	$\Delta G < 0$ always spontaneous	ΔG depends on T spontaneous if at high T
$\Delta S < 0$	ΔG depends on T spontaneous if at low T	$\Delta G > 0$ always nonspontaneous

$\Delta G < 0$ Spontaneous
$\Delta G > 0$ Nonspontaneous
$\Delta G = 0$ Equilibrium

Exergonic   Spontaneous

Energy released

Reactants

Products

Energy used

Products

Reactants

Nonspontaneous

Endergonic

Cold Pack

A single-use cold pack includes solid ammonium nitrate (NH_4NO_3), a soluble ionic compound, and water in a separate pouch. When the water pouch is torn, the water and ammonium nitrate mix to form an ionic solution. The reaction is endothermic, which means it absorbs heat from the surroundings, creating the cold feeling when touched.

$$NH_4NO_3\ (s) + heat \rightarrow NH_4^+\ (aq) + NO_3^-\ (aq)$$

The loss of heat from the surroundings decreases its entropy. However, the entropy of the system—the contents of the cold pack—increases because a solid compound is producing ions with much higher freedom, randomness, and disorder in solution. The increase in the system entropy more than compensates for the decrease in the entropy of the surroundings, keeping $\Delta G < 0$ and $\Delta S_{univ} > 0$.

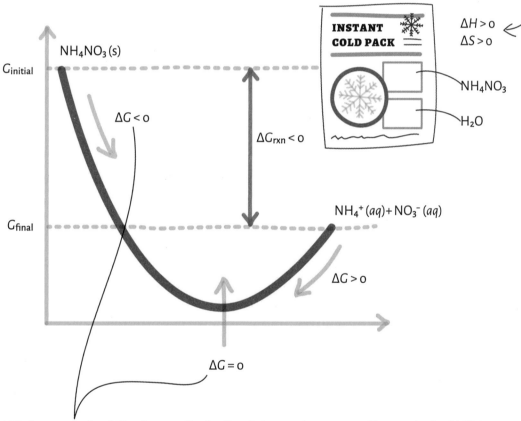

ΔG is the quantitative ability of the reaction to continue. As long as ΔG remains negative, the reaction progresses. As it does, the value of ΔG decreases, until it eventually reaches zero and establishes a dynamic equilibrium.

For the dissociation reaction in a cold pack, $\Delta H = +27$ kJ/mol and $\Delta S = +108.1$ J/mol·K. At ambient temperature and pressure, $\Delta G = -5.2$ kJ/mol for this exergonic reaction.

The magnitude of ΔG is large enough to make an instant cold pack efficient for approximately 15–20 minutes—long enough to treat a sprain or bruise.

THE ZEROTH AND THIRD LAWS

The **zeroth law of thermodynamics** deals with the direction of heat flow and thermal equilibrium between systems in physical contact, while the **third law of thermodynamics** provides the relationship between temperature and entropy of a substance. These two laws complement the principles of the first and second laws of thermodynamics.

The Zeroth Law

According to the zeroth law of thermodynamics, if two systems in physical contact are at different temperatures, q amount of heat will flow from the hot system to the cold one until the two reach the same temperature and are at thermal equilibrium.

Providing the theory behind the working principle of a thermometer is the zeroth law of thermodynamics. When a thermometer is placed in a sample, heat exchange occurs between them. This changes the density of the substance used in the thermometer, which in turn changes the temperature reading. When the thermometer and the sample reach thermal equilibrium, the thermometer reading is taken to be the temperature of the sample.

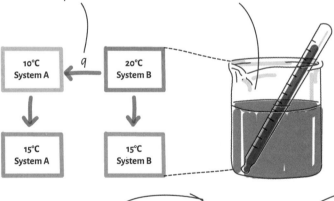

The Third Law

The third law of thermodynamics states that the entropy of a perfect crystalline substance is zero at absolute zero temperature (0 K).

Entropy is a measure of molecular randomness and disorder. When a substance is cooled to absolute zero temperature, all molecular motion ceases and a perfect crystalline structure forms; this highly ordered molecular arrangement represents zero entropy. Entropy starts to increase as the temperature of the substance increases. All entropy values reported in the scientific literature are called **absolute entropies** because they are determined based on the third law.

Entropy and temperature

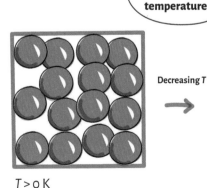

$T > 0\,K$
$S > 0$

Decreasing T

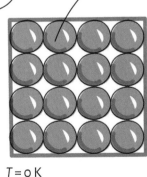

$T = 0\,K$
$S = 0$

EQUILIBRIUM THERMODYNAMICS

In the absence of outside interference, all spontaneous reactions proceed toward an equilibrium position. If the reaction species are not in their standard states, as is often the case, ΔG can still be determined. Gibbs energy can also provide thermodynamic information about the nature of the equilibrium position for a chemical reaction.

Gibbs Energy and Equilibrium

The **reaction quotient (Q)** is used to determine ΔG as a chemical reaction proceeds toward equilibrium under nonstandard conditions.

The equilibrium position is typically very close to the products for exergonic reactions because the negative value of ΔG drives the reaction closer to the product side.

The concentration difference between the reactants and products—expressed by the reaction quotient (Q)—is a driving force for the reaction in the forward direction, as long as $Q < K$ and $\Delta G < 0$.

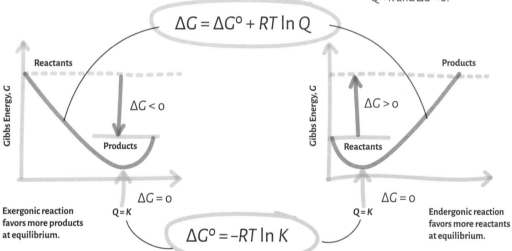

$$\Delta G = \Delta G^\circ + RT \ln Q$$

Reactants — Products

$\Delta G < 0$

Gibbs Energy, G

Products

$\Delta G = 0$

$Q = K$

Exergonic reaction favors more products at equilibrium.

Gibbs Energy, G

Products

$\Delta G > 0$

Reactants

$\Delta G = 0$

$Q = K$

Endergonic reaction favors more reactants at equilibrium.

$$\Delta G^\circ = -RT \ln K$$

When the reaction reaches equilibrium, $Q = K$ and $\Delta G = 0$. If $K \gg 1$, the equilibrium lies closer to the products and $\Delta G^\circ < 0$ at standard conditions.

If $K \ll 1$, then the equilibrium is closer to the reactants and $\Delta G^\circ > 0$ for the reaction. Endergonic reactions fall into this category—because the forward reaction is nonspontaneous, there are mostly reactants present at equilibrium.

ΔG° refers to the Gibbs energy change for the reaction when all species are at their standard states. It can be determined from Gibbs energy of formation data (ΔG_f°) widely available in the scientific literature.

$$\Delta G^\circ_{rxn} = \sum n \times \Delta G_f^\circ \text{ (products)} - \sum n \times \Delta G_f^\circ \text{ (reactants)}$$

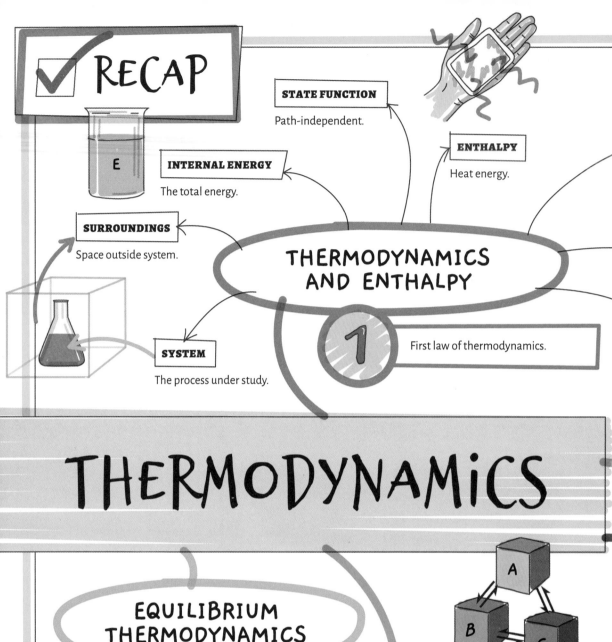

STATE FUNCTION

Path-independent.

INTERNAL ENERGY

The total energy.

ENTHALPY

Heat energy.

SURROUNDINGS

Space outside system.

THERMODYNAMICS AND ENTHALPY

SYSTEM

The process under study.

1 First law of thermodynamics.

THERMODYNAMICS

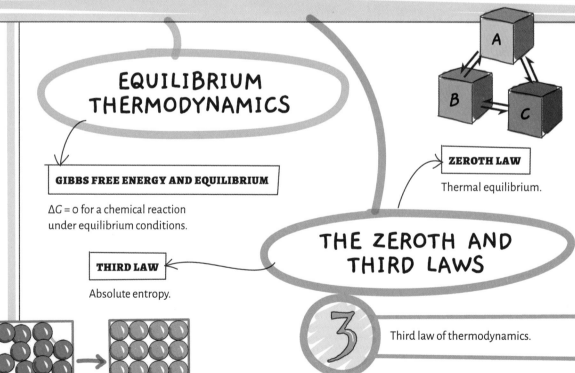

EQUILIBRIUM THERMODYNAMICS

GIBBS FREE ENERGY AND EQUILIBRIUM

$\Delta G = 0$ for a chemical reaction under equilibrium conditions.

THIRD LAW

Absolute entropy.

ZEROTH LAW

Thermal equilibrium.

THE ZEROTH AND THIRD LAWS

3 Third law of thermodynamics.

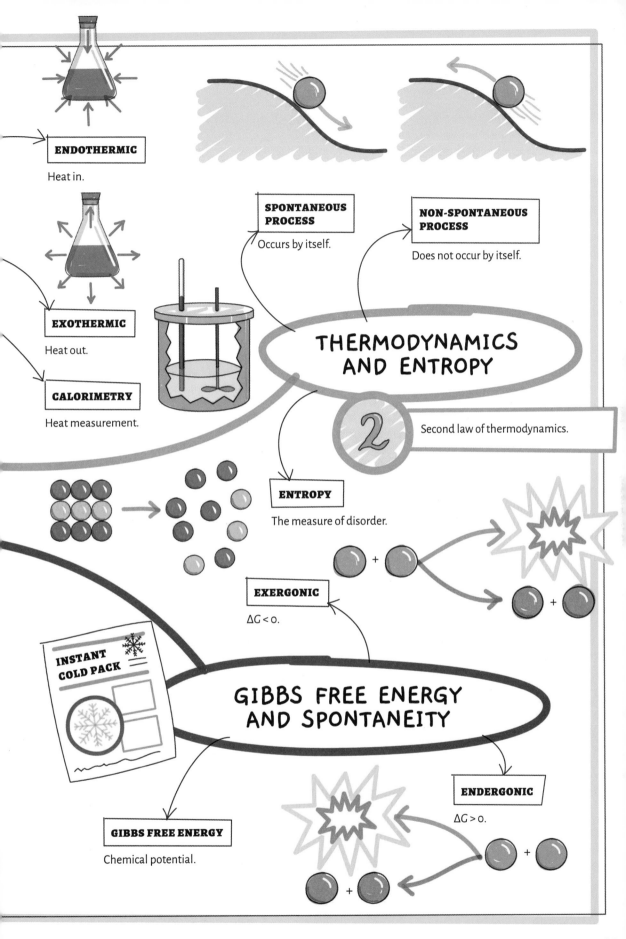

ENDOTHERMIC

Heat in.

EXOTHERMIC

Heat out.

CALORIMETRY

Heat measurement.

SPONTANEOUS PROCESS

Occurs by itself.

NON-SPONTANEOUS PROCESS

Does not occur by itself.

THERMODYNAMICS AND ENTROPY

2

Second law of thermodynamics.

ENTROPY

The measure of disorder.

EXERGONIC

$\Delta G < 0$.

INSTANT COLD PACK

GIBBS FREE ENERGY AND SPONTANEITY

GIBBS FREE ENERGY

Chemical potential.

ENDERGONIC

$\Delta G > 0$.

CHAPTER 14

ELECTROCHEMISTRY

Electrochemistry provides a scientific explanation for the interactions between electrical energy and chemical change, by studying chemical reactions that occur at the interface of an electrode and an electrolyte. This branch of chemistry involves oxidation-reduction reactions, in which electrons are transferred between certain reaction species. There are two main approaches: the use of spontaneous chemical reactions to produce electricity, and the use of electricity to bring about nonspontaneous chemical change. In both cases, the focus is the exchange of work (electrical power) between a system and its surroundings.

ELECTRONS iN ACTiON

When electric charge is in motion it generates electric current. This means the flow of negatively charged electrons through a medium such as a wire, or ions through an electrolyte solution, will create an electrical current. Redox reactions also have the potential to generate electric current because they involve the transfer of electrons from a substance of low electron affinity to a substance with higher electron affinity.

Redox Reactions

A substance with low electron affinity loses electrons and is **oxidized** in a redox reaction, while another substance with high electron affinity is **reduced** as it accepts those electrons.

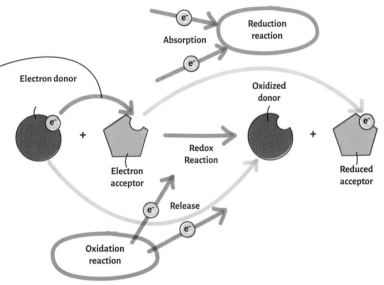

The electron transfer is fast when the two substances trading electrons are in direct contact, a configuration in which electrons are generated and consumed by the redox reaction without the production of useful electric power. If, however, the oxidized and reduced substances are separated and electrons are forced to move through a wire as the redox reaction takes place, an electric current can be generated for external use.

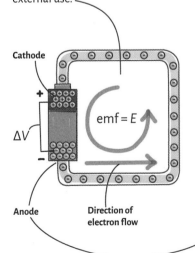

Electric Current

Electric current, which is measured in units of **amperes (A)**, is defined as the flow of electrical charge measured in **coulombs (C) per second**. An electron has a charge of 1.602×10^{-19} C, so 1 A of electric current corresponds to the flow of 6.242×10^{18} electrons per second.

In a redox reaction, metal **electrodes** connected by a conductive wire are used to transfer electrons from the oxidation side to the reduction side. The electrode where oxidation takes place is the **anode**, while the electrode where reduction occurs is the **cathode**.

Measured in **volts (V)**, the **potential energy difference (ΔV)** between the two electrodes forces the electron flow. This creates **direct current**, which is the electrical power generated by batteries.

Electromotive force (emf) is the potential difference when current flows between the anode and the cathode. In electrochemistry, electromotive force is commonly referred to as **cell potential (E)**.

VOLTAIC CELLS

A **voltaic cell** (also known as a **galvanic cell**) is an electrochemical cell in which a spontaneous oxidation-reduction reaction takes place, converting chemical potential energy into electrical energy to produce an electrical current. A voltaic cell has two compartments; oxidation takes place in one compartment, while reduction occurs in the other. This configuration establishes the foundation of how batteries operate.

Voltaic Cell Construction

In a voltaic electrochemical cell, oxidation takes place at the interface between a metal anode, such as zinc (Zn), and an aqueous electrolyte solution of the metal, such as zinc nitrate ($Zn(NO_3)_2$). Conventionally, the anode is located in the left compartment in the device setup.

Reduction takes place at the interface between a metal cathode, such as copper (Cu), and an aqueous solution of the metal, such as copper (II) nitrate ($Cu(NO_3)_2$).

The anode and cathode are connected via a conductive wire so electrons can move from the anode toward the cathode, supplying the necessary electrons for the redox reaction to proceed spontaneously.

The potential difference between the anode and the cathode forces the electrons to move, creating an electrical current that can be measured with a voltmeter. The amount of current generated depends on the potential energy difference during electron flow (the electromotive force or emf). A voltaic cell produces a positive cell potential (E) reading.

As the redox reaction proceeds spontaneously, the oxidation process in the left compartment will generate additional Zn^{2+} ions, while the reduction process consumes the available Cu^{2+} ions in the right compartment.

In order to maintain electrical charge neutrality, a **salt bridge** supplies additional anions and cations to the left and right compartments, respectively. Otherwise, the redox reaction does not proceed spontaneously and the electron flow ceases.

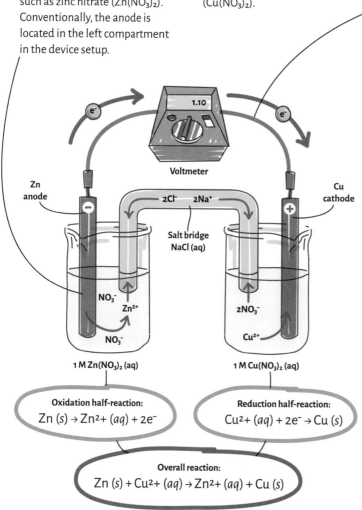

Zn anode

Cu cathode

1.10

Voltmeter

$2Cl^-$ $2Na^+$

Salt bridge
NaCl (aq)

NO_3^-
Zn^{2+}
NO_3^-

$2NO_3^-$
Cu^{2+}

1 M $Zn(NO_3)_2$ (aq)

1 M $Cu(NO_3)_2$ (aq)

Oxidation half-reaction:
$Zn\ (s) \rightarrow Zn^{2+}\ (aq) + 2e^-$

Reduction half-reaction:
$Cu^{2+}\ (aq) + 2e^- \rightarrow Cu\ (s)$

Overall reaction:
$Zn\ (s) + Cu^{2+}\ (aq) \rightarrow Zn^{2+}\ (aq) + Cu\ (s)$

Standard Cell Potential

When a voltaic cell operates under standard thermodynamic conditions, which means all the reactants and products are in their standard states (all solutions have a concentration of 1.0 M, and all gases are at 1 atm pressure), the measured cell potential is called the **standard cell potential ($E°_{cell}$)**. The temperature is typically assumed to be 25°C.

Cell potential depends on the relative tendencies of the reactants to undergo spontaneous oxidation and reduction. A substance with a high affinity for electrons combined with another substance exhibiting a low affinity for electrons produces a large positive cell potential. The larger the cell potential, the higher the tendency for the redox reaction to occur spontaneously.

Connect to other half-cell with various half-reactions to measure standard electrode potentials.

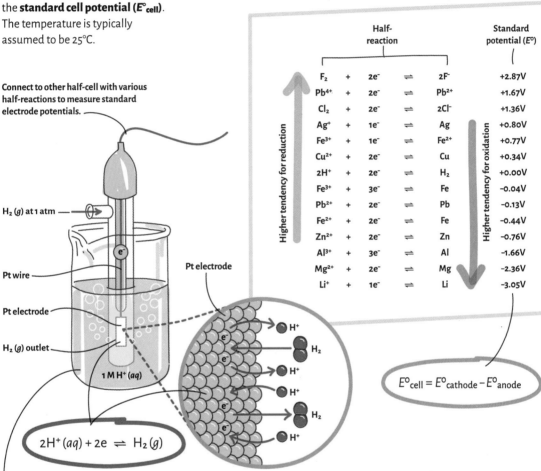

	Half-reaction			Standard potential ($E°$)
F_2	+	$2e^-$	$\rightleftharpoons$ $2F^-$	+2.87V
Pb^{4+}	+	$2e^-$	$\rightleftharpoons$ Pb^{2+}	+1.67V
Cl_2	+	$2e^-$	$\rightleftharpoons$ $2Cl^-$	+1.36V
Ag^+	+	$1e^-$	$\rightleftharpoons$ Ag	+0.80V
Fe^{3+}	+	$1e^-$	$\rightleftharpoons$ Fe^{2+}	+0.77V
Cu^{2+}	+	$2e^-$	$\rightleftharpoons$ Cu	+0.34V
$2H^+$	+	$2e^-$	$\rightleftharpoons$ H_2	+0.00V
Fe^{3+}	+	$3e^-$	$\rightleftharpoons$ Fe	−0.04V
Pb^{2+}	+	$2e^-$	$\rightleftharpoons$ Pb	−0.13V
Fe^{2+}	+	$2e^-$	$\rightleftharpoons$ Fe	−0.44V
Zn^{2+}	+	$2e^-$	$\rightleftharpoons$ Zn	−0.76V
Al^{3+}	+	$3e^-$	$\rightleftharpoons$ Al	−1.66V
Mg^{2+}	+	$2e^-$	$\rightleftharpoons$ Mg	−2.36V
Li^+	+	$1e^-$	$\rightleftharpoons$ Li	−3.05V

Higher tendency for reduction

Higher tendency for oxidation

H_2 (g) at 1 atm

Pt wire

Pt electrode

H_2 (g) outlet

Pt electrode

1 M H^+ (aq)

$$2H^+ (aq) + 2e \rightleftharpoons H_2 (g)$$

$$E°_{cell} = E°_{cathode} - E°_{anode}$$

Under standard conditions the anode and the cathode have their own **standard electrode potential ($E°$)**, which provides a measure of how strongly oxidation and reduction half-reactions tend to occur spontaneously.

The standard cell potential is the difference between the cathode and anode electrode potentials.

A **standard hydrogen electrode (SHE)** is constructed with a platinum (Pt) electrode immersed in 1 M strong acid solution, where hydrogen ions (H^+) are reduced to H_2 gas. By convention, the standard electrode potential for SHE is taken to be 0.0 V. All other standard electrode potentials are measured against SHE, and are usually tabulated as standard reduction potentials.

Half-reactions with a higher standard reduction potential (more positive) compared to the SHE tend to occur at the cathode as reduction half-reactions.

Half-reactions with a lower standard reduction potential (more negative) than the SHE tend to take place at the anode as oxidation half-reactions.

GiBBS FREE ENERGY AND ELECTROCHEMiSTRY

A voltaic cell is designed to produce electricity, which requires the use of a spontaneous redox reaction that is capable of generating a positive cell potential. As Gibbs free energy provides the criterion for spontaneity, cell potential and ΔG are related.

Reaction at standard-state conditions.

Spontaneity at Standard Conditions

As long as $\Delta G° < 0$, a positive cell potential ($E°_{cell} > 0$) will be generated, because the reaction proceeds spontaneously under standard conditions. **Faraday's constant (F)**—the charge in coulombs for 1 mole of electrons —provides the mathematical relationship between $E°_{cell}$ and $\Delta G°$.

$\Delta G°$ is related to the equilibrium constant (K) of the redox reaction taking place under standard conditions.

	spontaneous	at equilibrium	non-spontaneous
$\Delta G°$	< 0	0	> 0
K	>1	1	<1
$E°_{cell}$	>0	0	<0

$\Delta G°$

$\Delta G° = -nFE°_{cell}$

$\Delta G° = -RT \ln K$

$E°_{cell}$

K

$$E°_{cell} = \frac{-RT}{nF} \ln K$$

$R = 8.314 \ \dfrac{J}{mol.K}$

Moles of electrons

Faraday's constant = 96485 C/mol

The $E°_{cell}$ generated in a voltaic cell and the equilibrium constant for the redox reaction taking place in the cell are naturally related to each other mathematically.

The mathematical relationships between $E°_{cell}$, $\Delta G°$, and K provide the conditions under which a voltaic cell must operate in order to generate electricity.

A battery is a voltaic cell generating electricity. Discharging a battery means $\Delta G° < 0$ and $E°_{cell} > 0$; the battery continues to operate until the redox reaction reaches equilibrium. With a rechargeable battery, electricity is used to reverse the redox reaction and restore its original state, so the battery can operate again.

Spontaneity at Nonstandard Conditions

For a voltaic cell constructed with zinc and copper electrodes $E°_{cell}$ = 1.10 V under standard conditions, when the electrolyte solutions involved have a concentration of 1.0 M. However, the same voltaic cell will produce 1.17 V of cell potential (E_{cell}) when the concentrations of the electrolyte solutions are 0.01 M at the anode and 2.0 M at the cathode.

For the spontaneous redox reaction taking place in a voltaic cell operating under nonstandard conditions, the Gibbs free energy (ΔG) is given by the **Nernst equation**, where Q is the reaction quotient for the overall redox reaction.

$$\Delta G = \Delta G° + RT \ln Q$$

Different cell potential under nonstandard-state conditions.

Voltmeter

1.17

e^- e^-

Zn anode   Cu cathode

2Cl⁻   2Na⁺

Salt bridge
NaCl (aq)

NO_3^-
Zn^{2+}
NO_3^-

$2NO_3^-$
Cu^{2+}

$Zn (s) \rightarrow Zn^{2+} (aq) + 2e^-$ $Cu^{2+} (aq) + 2e^- \rightarrow Cu (s)$

Nonstandard electrolyte concentrations.

0.01 M $Zn(NO_3)_2$ (aq)   2.0 M $Cu(NO_3)_2$ (aq)

$$E_{cell} = E°_{cell} - \frac{RT}{nF} \ln Q$$

The cell potential (E_{cell}) measured under nonstandard states is related to the standard cell potential ($E°_{cell}$).

For a voltaic cell that is designed to generate electricity, the redox reaction must occur spontaneously. This naturally means that as long as $\Delta G < 0$, the cell continuous to operate with $E_{cell} > 0$ until the reaction reaches equilibrium, when $\Delta G = 0$.

BATTERIES AND FUEL CELLS

A battery utilizes a spontaneous redox reaction to not only generate electricity, but also to store it. Batteries are contained, compact units constructed with the two-electrode configuration of a voltaic cell, and electricity is only produced when the battery is in use. Fuel cells, on the other hand, work on a different principle, which requires a continuous supply of reactants involved in the redox reaction. These cells convert chemical energy to electrical energy, but do not store it.

Batteries

There are two general types of battery: **primary batteries**, which are single use, and **secondary batteries**, which are rechargeable and designed to be used multiple times.

Alkaline battery

The most common type of single-use battery is an alkaline battery, which gets its name because a base, potassium hydroxide (KOH), is used as an electrolyte. This type of battery should be recycled carefully so that the strong base does not leak into the environment.

Zinc (Zn) metal is used for the casing of an alkaline battery, and this also serves as the anode. A graphite rod immersed in an electrolyte paste consisting of manganese (IV) oxide (MnO_2) and potassium hydroxide (KOH) is the cathode of the battery.

$$2MnO_2\ (s) + 2H_2O\ (l) + 2e^- \rightarrow 2MnO(OH)\ (s) + 2OH^-\ (aq)$$

Manganese is reduced at the cathode, consuming the electrons supplied by the anode.

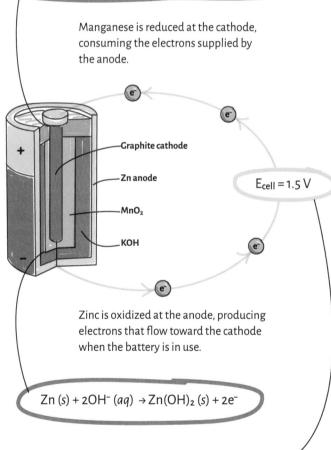

Graphite cathode

Zn anode

MnO_2

KOH

$E_{cell} = 1.5\ V$

Zinc is oxidized at the anode, producing electrons that flow toward the cathode when the battery is in use.

$$Zn\ (s) + 2OH^-\ (aq) \rightarrow Zn(OH)_2\ (s) + 2e^-$$

A typical alkaline battery generates 1.5 V of cell potential, as long as the electrodes have ions available. When one of the electrodes is depleted (runs out of ions), the battery is dead.

Fuel Cells

Like batteries, fuel cells also produce electricity by utilizing a redox reaction, but reactants must be supplied continuously. The most common fuel cell is the hydrogen fuel cell used in space shuttles.

In a hydrogen fuel cell, gaseous hydrogen is oxidized to H^+, generating electrons that are allowed to flow toward the cathode. To speed up the oxidation process, a platinum (Pt) catalyst is used at the anode.

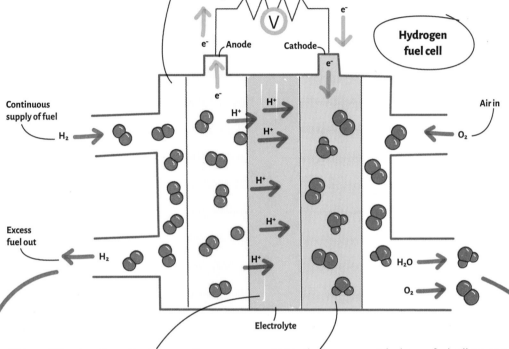

0.5–0.8 V per cell

Hydrogen fuel cell

Continuous supply of fuel

H_2

Anode   Cathode

Air in

O_2

Excess fuel out

H_2

H_2O

O_2

Electrolyte

H^+ ions diffuse into the cathode through an electrolyte medium, which is an aqueous potassium hydroxide (KOH) solution.

Gaseous oxygen (O_2) in the air reacts with H^+ ions at the cathode to produce water.

A hydrogen fuel cell generates 0.5–0.8 V of cell potential, but numerous fuel cells can be connected in series to increase the power output.

$$H_2\,(g) \rightarrow 2H^+\,(aq) + 2e^-$$

$$1/2O_2\,(g) + 2H^+\,(aq) + 2e^- \rightarrow H_2O\,(l)$$

$$2H_2\,(g) + O_2\,(g) \rightarrow 2H_2O\,(l)$$

The redox reaction in the fuel cell produces water as the product. In a space shuttle, astronauts drink this water.

It is projected that hydrogen fuel cells will one day replace the current conventional power generation methods for transportation and household use. However, the widespread commercial use of hydrogen fuel cells first requires the development of readily available hydrogen sources and cheaper catalyst materials.

ELECTROLYTIC CELLS

When a spontaneous reaction takes place in a voltaic cell, electrical power is generated. Electrolysis is the process that takes place in an electrolytic cell to which an external electrical current is supplied in order to drive an otherwise non-spontaneous redox reaction.

Driving Nonspontaneous Reactions

A voltaic cell constructed with a cadmium (Cd) anode and a copper (Cu) cathode generates 0.74 V of current under standard conditions, because the overall redox reaction in which Cd is oxidized is spontaneous with $\Delta G° < 0$. The reverse reaction is nonspontaneous and will not occur under standard conditions.

If a power supply is used to add electric current greater than 0.74 V to the Cd-Cu cell, the electron flow can be reversed, which means the nonspontaneous reaction in the voltaic cell now becomes spontaneous. The new electrochemical cell construction is called an electrolytic cell.

The anode in the voltaic cell becomes the cathode in the electrolytic cell; copper is now oxidized at the anode and cadmium is reduced at the cathode. The electrons, which are supplied externally and are not generated by the redox reaction, flow toward the cathode in this configuration.

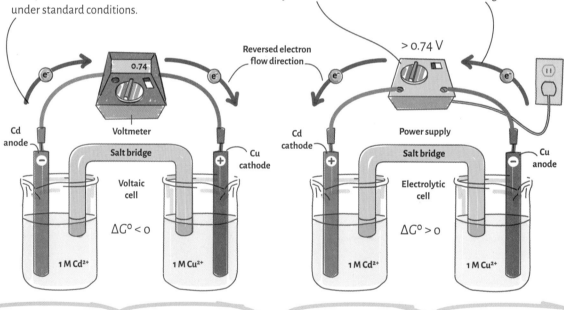

Cd (s) → Cd²⁺ (aq) + 2e⁻   Cu²⁺ (aq) + 2e⁻ → Cu (s)

Cd (s) + Cu²⁺ (aq) → Cd²⁺ + Cu (s)

Cu² (aq) + 2e⁻ → Cd (s)   Cu (s) → Cu²⁺ (aq) + 2e⁻

Cd²⁺ (aq) + Cu (s) → Cd (s) + Cu²⁺ (aq)

For an electrolytic cell, $E°_{cell} < 0$, and no electric current will be generated because the overall redox reaction is non-spontaneous.

Therefore, an electrolytic cell must be supplied externally with the necessary electrons for the reaction to occur. The purpose of an

electrolytic cell is not to generate power, but to use electric power to carry out commercially important electrolysis processes.

Electrolysis

Electrolysis is a process in which electric current is used to drive otherwise non-spontaneous reactions. The reaction of hydrogen with oxygen to form water is spontaneous, which makes it possible to use the reaction in a fuel cell to generate electricity. However, by supplying electrical current in an electrolytic cell the reaction can be reversed, decomposing water into hydrogen and oxygen gas.

Water is oxidized in the anode to form oxygen gas using an external supply of electrons.

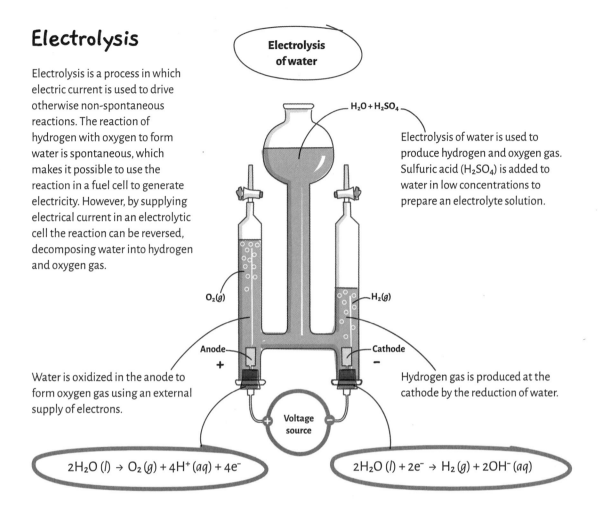

Electrolysis of water

$H_2O + H_2SO_4$

Electrolysis of water is used to produce hydrogen and oxygen gas. Sulfuric acid (H_2SO_4) is added to water in low concentrations to prepare an electrolyte solution.

$O_2(g)$

$H_2(g)$

Anode +

Cathode −

Voltage source

Hydrogen gas is produced at the cathode by the reduction of water.

$$2H_2O\,(l) \rightarrow O_2\,(g) + 4H^+\,(aq) + 4e^-$$

$$2H_2O\,(l) + 2e^- \rightarrow H_2\,(g) + 2OH^-\,(aq)$$

Electroplating

An important industrial application of electrolysis is **electroplating**, in which metal surfaces are uniformly plated with another metal. This process would not occur spontaneously.

In an electrolytic cell, silver from an aqueous solution of silver ions (Ag^+) can be plated onto a metallic object, such as a piece of cutlery.

With the supply of electrons, silver metal is oxidized at the anode, which produces silver ions. The object being plated is held in the same electrolyte solution and serves as the cathode. The silver ions get reduced to metallic silver, which deposits on the surface of the object.

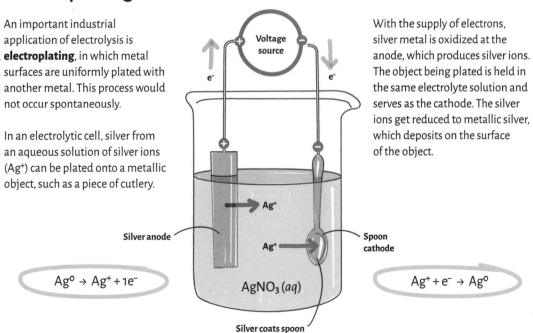

Voltage source

e^-

e^-

Ag^+

Silver anode

Ag^+

Spoon cathode

$AgNO_3\,(aq)$

Silver coats spoon

$$Ag^o \rightarrow Ag^+ + 1e^-$$

$$Ag^+ + e^- \rightarrow Ag^o$$

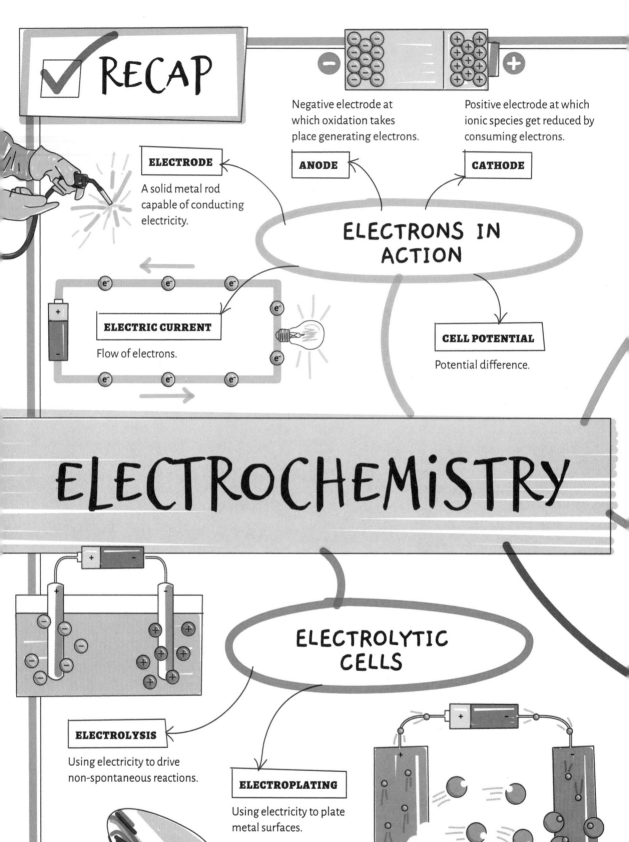

RECAP

Negative electrode at which oxidation takes place generating electrons.

Positive electrode at which ionic species get reduced by consuming electrons.

ELECTRODE

A solid metal rod capable of conducting electricity.

ANODE

CATHODE

ELECTRONS IN ACTION

ELECTRIC CURRENT

Flow of electrons.

CELL POTENTIAL

Potential difference.

ELECTROCHEMISTRY

ELECTROLYTIC CELLS

ELECTROLYSIS

Using electricity to drive non-spontaneous reactions.

ELECTROPLATING

Using electricity to plate metal surfaces.

ELECTROCHEMICAL CELL

A two-compartment configuration with separated anode and cathode electrodes.

SALT BRIDGE

Supplies ions into the anode and cathode of an electrochemical cell.

STANDARD CELL POTENTIAL

Current under standard conditions.

VOLTAIC CELLS

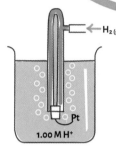

STANDARD HYDROGEN ELECTRODE

Assigned a zero half-cell potential.

STANDARD ELECTRODE POTENTIAL

Measured electrical current for an electrochemical cell operating under standard conditions.

NERNST EQUATION

Electrochemical cell potential under nonstandard conditions.

$$E_{cell} = E^o{}_{cell} - \frac{RT}{nF} \ln Q$$

FARADAY'S CONSTANT

Electric charge carried by 1 mole of electrons = 96485 C/mol.

GIBBS FREE ENERGY AND ELECTROCHEMISTRY

PRIMARY BATTERIES

Single use because the redox reaction involved is nonreversable.

SECONDARY BATTERIES

Rechargeable with reversible redox reaction.

BATTERIES AND FUEL CELLS

FUEL CELLS

Power generated continuously as long as reactants are supplied into the cell.

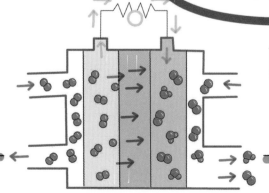

GLOSSARY

Citric acid makes lemons taste sour.

ACID
A substance that donates protons to another substance, giving aqueous solutions a sour taste and a pH value less than 7.

ACID RAIN
Rainwater with a pH of less than 5.6. Caused by polluting gases in the air creating acidic conditions when mixed with water.

ACIDIC ANHYDRIDES
The oxides of some nonmetals, such as carbon, sulfur, and nitrogen, which are typically found in gaseous form in polluted air. Form acidic conditions when mixed with rainwater.

ALPHA PARTICLE
A positively charged helium atom emitted by an unstable nuclei during nuclear changes.

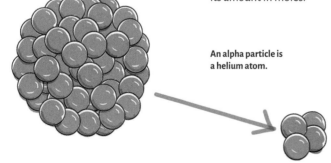

An alpha particle is a helium atom.

AMPHOTERIC SUBSTANCE
A substance, such as water, that can act both as a base and an acid by accepting or donating protons in an aqueous solution.

ANODE
The electrode at which oxidation takes place generating electrons in an electrochemical cell.

ATOM
The basic building block of all matter. Composed of a nucleus that houses protons and neutrons, with electrons located in the largely empty space around it.

ATOMIC MASS
The weighted mass average of all known isotopes of an element, based on their abundance in nature, which defines its accepted average mass. Expressed as an atomic mass unit, or amu.

ATOMIC NUMBER
The identifying and characteristic property of an element's position in the periodic table defined as the number of protons in its nucleus.

AVOGADRO'S LAW
A mathematical expression that provides the direct relationship between the volume of a gas and its amount in moles.

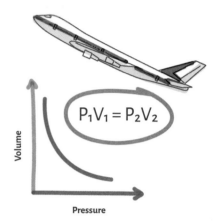

$$P_1V_1 = P_2V_2$$

Boyle's law: pressure versus volume.

BASE
A substance that accepts protons from another substance, giving aqueous solutions a bitter taste and a pH value greater than 7.

BETA PARTICLE
A negatively charged electron emitted by an unstable nuclei during nuclear changes.

BOYLE'S LAW
A mathematical expression providing the inverse relationship between the pressure and volume of a gas sample.

BROWNIAN MOTION
The completely random and ceaseless motion of particles in a gas sample.

BUFFER
An aqueous solution, made of a weak acid or base and its salt, which is capable of resisting changes in pH when strong acids or bases are added to it.

CALORIMETRY
An experimental technique utilized by scientists to measure the exact amount of heat energy transferred between a system and its surroundings during a physical or chemical change.

CATHODE
The electrode at which reduction takes place consuming electrons in an electrochemical cell.

CHARLES'S LAW
A mathematical expression providing the direct relationship between the temperature and volume of a gas sample.

CHEMICAL EQUILIBRIUM
A specific state of a chemical reaction, reached after a certain time into the reaction, which is characterized by equal rates in the forward and reverse directions. When this state is reached, there is no net change in the concentrations of the products and reactants.

CHEMICAL PROPERTY
A property of matter that can be measured or observed when it chemically changes into a different substance involving the same elements.

Charles's law: volume versus temperature.

CHEMISTRY
A fundamental branch of science building knowledge about matter; its composition, structure, and changes, as well as its interaction with other matter and energy.

COMPOUND
A pure form of matter composed of two or more elements that are combined chemically in a specific way according to the law of definite proportions.

COVALENT BOND
A form of chemical bonding that forms between nonmetal atoms of similar electronegativity values through the sharing of valence electrons, in order to satisfy the octet rule.

DENSITY
A characteristic property of matter, defined as mass per unit volume.

ELASTIC COLLISION
Collision between gas particles, during which energy can be exchanged, but the overall energy is conserved.

ELECTROCHEMISTRY
A branch of chemistry that studies the interconversion of chemical and electrical energies during oxidation and reduction reactions occurring at the interface between an electrode and an electrolyte solution.

ELECTRODE
An electrical conductor—usually a metal—that is used to transfer electrons from the anode to the cathode in an electrochemical cell.

ELECTROLYTE
A substance that ionizes when dissolved in water, such as table salt, forming a solution that conducts electricity.

ELECTROMOTIVE FORCE
The potential difference between the anode and the cathode of an electrochemical cell during the flow of electric current.

Electrolytes are crucial for human health.

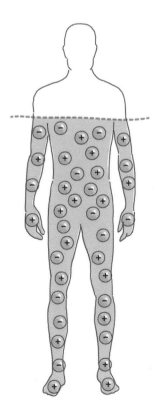

Constant pressure calorimeter for enthalpy measurements.

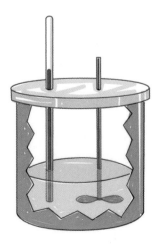

ELECTRONEGATIVITY

The ability of an atom to attract electrons to itself. Values ranging from 0.7–4.0 for the elements in the periodic table.

ENDOTHERMIC PROCESS

A physical or chemical process with a positive enthalpy change value, which requires an input of heat energy to occur.

ENTHALPY

Heat energy transferred during a physical or chemical process that takes place in a system and its surroundings.

ENTROPY

A thermodynamic quantity that provides a measure of how spread out or dispersed the energy of a system is. Often associated with the randomness and freedom of the particles in the system.

EQUILIBRIUM CONSTANT

The fixed ratio of product and reactant concentrations under conditions of dynamic equilibrium for a reversible chemical reaction at constant temperature according to the law of mass action.

Gay-Lussac's law: Pressure versus temperature.

EXOTHERMIC PROCESS

A physical or chemical process with a negative enthalpy change value, which releases heat energy as it takes place.

FARADAY'S CONSTANT

The charge, in coulombs (C), carried by one mole of electrons, has a numerical value of 96485 C/mol.

GALVANIC CELL

A two-cell construction of an electrochemical cell, such as a battery, in which a spontaneous oxidation-reduction reaction takes place producing electrical current. Also known as a voltaic cell.

GAMMA PARTICLE

A highly energetic photon with no mass or charge, emitted by unstable nuclei. These particles are capable of penetrating human skin and cause cell damage.

GAY-LUSSAC'S LAW

A mathematical expression providing a direct relationship between the temperature and pressure of a gas sample.

GIBBS FREE ENERGY

Chemical potential energy that provides a quantitative criterion for the direction of spontaneous change as well as for the ability of a physical or chemical process to bring about change.

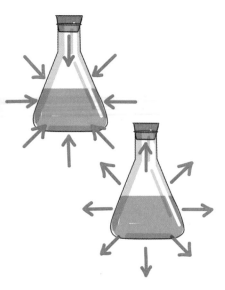

Endothermic and exothermic processes showing the direction of heat flow.

ELECTRON

A negatively charged subatomic particle located in the empty space around the nucleus of an atom. Electrons are largely responsible for observed chemical properties.

ELECTRON AFFINITY

A measure of how easily an atom can accept an electron, which provides strong evidence about the chemical bonding properties of an element.

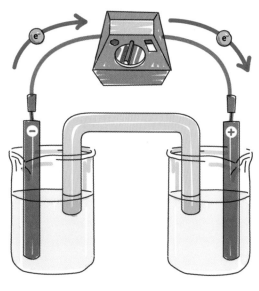

A galvanic cell construction visualizing how batteries operate.

HALF-LIFE

A useful timescale that provides information about how long it takes half of a radioactive substance to decay into a more stable form.

HYDROGEN BONDING

An attractive intermolecular force that exists between molecules with an F, O, or N atom directly bonded to a H atom.

IDEAL GAS LAW

A mathematical expression that provides the relationships between the four identifying properties of a gas sample: pressure, temperature, volume, and amount in moles.

INTERMOLECULAR FORCE

An attractive force, whose strength depends on molecular polarity. The intermolecular force keeps molecules together in covalently bonded substances.

IONIC BOND

A strong electrostatic attraction between oppositely charged ions, resulting from the complete transfer of valence electrons between a metal and nonmetal atom.

Hydrogen

Helium

Carbon

Line spectra showing quantized energies possessed by electrons.

IONIZING RADIATION

Electromagnetic radiation (light) that is capable of removing electrons from atoms and molecules, causing tissue damage in biological organisms.

ISOTOPE

Atoms of the same element with an identical number of protons in their nuclei, but a different number of neutrons.

LAW OF MASS ACTION

Provides the mathematical definition of the equilibrium constant for a reversible chemical reaction, as the fixed ratio of product and reactant concentrations at constant temperature.

LINE SPECTRUM

Characteristic spectral lines of distinct color and energy that are formed when light emitted or absorbed by an element is visualized through the use of a spectrometer.

MATTER

Anything made of atoms in the physical universe that occupies space and has a rest mass with a specific amount of energy.

MASS

A measure of the amount of matter that an object contains, regardless of its location in the universe. Specified with the SI unit of kilogram.

METALLIC BOND

Forces that keep metal atoms together via electrostatic attraction between positively charged metal nuclei and a sea of freely moving electrons.

MIXTURE

Matter with a variable composition, consisting of two or more components mixed together, but separable by means of physical methods, such as filtration, distillation, and vaporization.

MOLAR MASS

The mass, in grams, of one mole of a substance.

Free electron movement around atomic nuclei forms the basis of metallic bonding.

Dipole-dipole intermolecular forces between polar molecules.

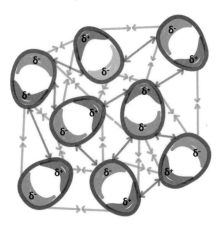

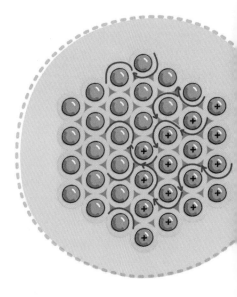

MOLARITY

A concentration term defined as the amount of substance (in moles) dissolved in one liter of solution. Useful for specifying how many particles are present in a homogeneous mixture or solution.

MOLE

A convenient SI unit of measurement representing 6.022×10^{23} number of particles in a given sample (Avogadro's number). Allows chemists to determine the exact numbers of atoms and molecules needed to prepare samples, or to carry out chemical reactions.

MOLECULE

A chemical unit formed by two or more permanently combined atoms with similar electronegativities through the formation of covalent bonds that result from the sharing of valence electrons.

NEUTRALIZATION

The chemical reaction between an acid and a base that forms water and an ionic compound as the reaction products.

NEUTRON

A neutral subatomic particle with a mass of 1.00866 amu (atomic mass units), located in the nucleus of an atom.

Nuclear binding energy keeps the nucleus intact.

NUCLEAR BINDING ENERGY

The energy required to separate an atom's nucleus into its protons and neutrons; essentially represents the energy that keeps the nucleus together.

NUCLEAR FORCE

A very strong attractive force that acts between subatomic particles that are extremely close to each other, such as protons and neutrons.

NUCLEUS

The heavy core of an atom that houses protons and neutrons, essentially providing the entire mass of the atom.

OCTET RULE

The tendency of an atom to acquire 8 electrons in its outermost energy shell when forming chemical bonds with other atoms by gaining, losing, or sharing valence electrons.

ORBITAL

A three-dimensional charge-cloud region around the nucleus with discrete energy that represents a 90 percent probability that an electron can be found in its well-defined volume element in space.

OXIDATION

The loss of electrons by a substance during a chemical reaction.

PERIODIC TABLE

Arrangement of the elements based on their atomic number and periodic changes of chemical properties.

pH

Power of hydrogen. Mathematically defined as the –log of the hydrogen ion concentration in an aqueous solution, providing a numerical scale for the acidic or basic character of a substance.

pH INDICATOR

A complex organic molecule with a weak acidic or alkaline character and limited ionization ability in water, which is capable of displaying a distinct color in varying pH environments.

PHOTONS

Massless particles of light composed of tiny packets of electromagnetic energy that travel in space in wave form, at the speed of light when in a vacuum.

PHYSICAL PROPERTY

A property of matter that can be measured or observed without changing its identity, such as melting ice, the smell of a rose, or the color of the ocean.

Three-dimensional representation of an atomic orbital.

The molarity concept makes solution preparation convenient for chemists.

Solubility of solids in liquids increases with temperature.

PROTON
A positively charged subatomic particle with a mass of 1.00728 amu (atomic mass units), located in the nucleus of an atom.

PRESSURE
The total force exerted by gas particles upon collision with a unit area of their container walls.

QUANTUM NUMBER
A number that defines the specific energy, shape, and other properties of orbitals.

REDUCTION
The gain of electrons by a substance during a chemical reaction.

SI UNIT
Abbreviation of the French Système international (d'unités), or International System of Units. Consists of seven standardized base units that are used to measure the amount of a substance (mole), temperature (kelvin), mass (kilogram), length (meter), electric current (ampere), luminosity (candela), and time (second).

SOLUBILITY
The maximum amount of substance that can dissolve in a solvent at a given temperature; dependent on the chemical nature of the substance and the solvent, as well as the temperature.

SOLUTION
A homogeneous mixture of two or more substances with a uniform composition throughout.

SPONTANEOUS PROCESS
A physical, chemical, or nuclear change that, once initiated, proceeds by itself without constant external interference.

SUBSTANCE
A pure form of matter with a well-defined and fixed composition.

VALENCE ELECTRONS
Electrons located in the outermost energy shell of an atom, which are largely responsible for the chemical properties of the elements.

VOLTAIC CELL
see Galvanic cell.

VSEPR THEORY
Valence shell electron pair repulsion (VSEPR) theory describes molecular shapes based on the electrostatic repulsion between the negatively charged valence electrons, as well as the polarity of the resulting molecular geometry.

Light scattered by dust particles in air according to the Tyndall effect.

THERMODYNAMICS
A branch of science that studies the energy changes accompanying physical and chemical processes, as well as the spontaneous nature of physical and chemical changes under a set of given environmental conditions.

TITRATION
A quantitative analysis technique used to find the exact concentration of an acidic or a basic substance in a sample by performing a neutralization reaction.

TYNDALL EFFECT
The scattering of a light beam by nanometer-sized particles suspended in a colloid mixture or air with random motion.

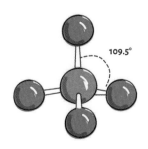

Tetrahedral molecular geometry is based on the VSEPR theory.

iNDEX

Acknowledgments

I would like to recognize some awesome people
who made this book a reality. Many thanks to the publisher
of UniPress Books, Nigel Browning, for taking a chance
on me as the author for this project. Special thanks to
Natalia Price–Cabrera, project manager and editor, for her
guidance, constant feedback, and patience from the start
to the completion of the book. Lindsey Johns deserves credit for
the brilliant design and powerful illustrations that make this
book a truly visual experience. I would also like to extend my
gratitude to my parents, Faik and Hatice, for always believing
in me. I now know writing a book is a difficult task, but also a
rewarding one. It would not have been possible to realize
Visual Learning Chemistry without the incredible hard work
by everyone involved. Thank you!